Student Study Guide/ Solutions Manual

for use with

Biochemistry

The Molecular Basis of Life

Third Edition

Dr. Trudy McKee

Dr. James R. McKee
University of the Sciences in Philadelphia

Prepared by
Dr. Patricia DePra
Westfield State College, Westfield, MA

Boston   Burr Ridge, IL   Dubuque, IA   Madison, WI   New York   San Francisco   St. Louis
Bangkok   Bogotá   Caracas   Lisbon   London   Madrid
Mexico City   Milan   New Delhi   Seoul   Singapore   Sydney   Taipei   Toronto

McGraw-Hill Higher Education

A Division of The McGraw-Hill Companies

Student Study Guide/Solutions Manual for use with
BIOCHEMISTRY: THE MOLECULAR BASIS OF LIFE, THIRD EDITION.
TRUDY MCKEE AND JAMES R. MCKEE.

Published by McGraw-Hill Higher Education, an imprint of The McGraw-Hill Companies, Inc.,
1221 Avenue of the Americas, New York, NY 10020. Copyright © The McGraw-Hill Companies,
Inc., 2003, 1999, 1996. All rights reserved.

2 3 4 5 6 7 8 9 0 BKM BKM 0 9 8 7 6 5 4 3 2

ISBN 0-07-242449-4

www.mhhe.com

STUDENT STUDY GUIDE/SOLUTIONS MANUAL

TABLE OF CONTENTS

ACKNOWLEDGEMENTS

Special thanks go to my biochemistry students who had the courage to ask questions, and were not afraid to bare their confusion for the sake of understanding. Their enthusiasm inspired me to create new techniques to explain key material and to augment the text. The joy of their "Aha!" moments continually motivated me, and made the teaching process fun.

I also thank the text and solutions authors, Drs. Trudy and James McKee, Dr. Michael Kalafatis for his helpful comments, and Dr. Bruce Morimoto, who prepared the previous edition of this study guide.

My deepest thanks go to my dear husband, Larry Springsteen, and to my family. Their love fills my heart and makes me whole.

Patricia A. DePra, March 2002

How to Succeed in Biochemistry

Biochemistry weaves together a study of the structure of biological molecules and their functions in living systems. How these biological molecules act, react, and interact contributes to the flow of energy that is life. Biochemistry integrates such diverse subjects as organic chemistry, physiology, and even physical chemistry, and takes them all a step further. Many students find that concepts that were a bear in previous courses are more interesting and easier to understand when placed in the context of how they affect life. Some of the most fundamental concepts (such as polarity and hydrogen bonding) help to explain why a particular molecule behaves the way it does.

Living organisms continually experience a flow of energy based on biochemical reactions. Your body continually metabolizes nutrients, taps into or adds to energy stores, produces heat, maintains the correct amount of water and ions, and catalyzes seemingly impossible chemical reactions so that you can think, breathe, move, eat, burp, you name it. At the end of this course, you'll have a deeper understanding of how chemistry plays an integral role in the functioning and energetics of living systems.

HOW TO SUCCEED IN BIOCHEMISTRY

- Spend time on biochemistry *every day*. Schedule it in, make it part of your routine.

- Read the text *before* it's discussed in class.

- Attend class and participate in the discussions.

- Do the homework problems.

- Keep up with any necessary memorization. Strive to *learn* rather than to memorize. Put each subject into a context of why each topic is important to life – this will help you to understand it, and understanding makes the learning (even the memorization) easier.

- Don't let tough sections slip by. Much of this material builds upon itself, so if you hit a rough patch, work hard to get through it with understanding.

- If concepts from your general and organic chemistry backgrounds seem a bit fuzzy, be sure to tackle them early. Many of them become recurring themes in later chapters.

- *ASK QUESTIONS and SEEK THE ANSWERS*. Leverage your professor's or teaching assistant's office hours, students who have had the course previously, and your fellow students.

To the memory of my daughter Emma Marie, who lived from December 24 - 26, 2001, and to the conscientious doctors and nurses at the Children's Hospital of Buffalo. Their expertise in biochemistry kept our little Emma alive against all logic, and their caring warmth was a real comfort to us.

May you find something in the study of this course that helps you to make another life more beautiful and more complete. May you find joy in learning the fascinating complexity of reactions and interactions between biomolecules, and with your new knowledge, may you grow to treasure life more each day.

CHAPTER ONE

Biochemistry: An Introduction

INTRODUCTION

What is life? complex and dynamic, organized and self-sustaining, cellular, information-based, able to adapt and evolve

Terms: biomolecules   metabolism   emergent properties   genes

macromolecules   homeostasis   enzymes   mutations

Domains: Bacteria, Archaea, Eukarya

THE LIVING WORLD

Prokaryotic (Bacteria and Archaea) vs. Eukaryotic Cells

Prokaryotes	vs.	Eukaryotes
no nucleus		nucleus contains genetic material
single-celled organisms		cells contain complex organelles
much smaller		many are multicellular

Bacteria

Archaea: includes extremophiles (contain extremozymes)

Eukarya

ADVANTAGES OF MULTICELLULAR OVER SINGLE-CELLED ORGANISMS:

1. provides a relatively stable environment for most of the organism's cells

2. capable for greater complexity in the organism's form and function

3. able to exploit environmental resources more effectively

Viruses

parasites that need a host to survive; can cause disease, can be agents of genetic change via transduction (transfer of genetic material from one host cell to another)

BIOMOLECULES

Organic functional groups:

amine   carboxylic acid   amide   thiol (sulfhydryl)

alcohol   aldehyde   ketone

alkene (double bond)   ester

Hydrophobic vs. hydrophilic

α carbon = the carbon *adjacent to* a carbonyl carbon

Major Classes of Small Biomolecules

These <u>building blocks</u> are components of these <u>larger biomolecules</u>:

amino acids	peptides, polypeptides, proteins
sugars	carbohydrates
fatty acids	lipids
nucleotides	nucleic acids (RNA, DNA)

FUNCTIONS OF SMALL BIOMOLECULES:

- to synthesize larger molecules,

- to carry out special biological functions (e.g., as neurotransmitters or hormones), and/or

- to take part in complex reaction pathways

AMINO ACIDS, PEPTIDES, AND PROTEINS

- Amino acids contain an amino group, a carboxylic acid group, and a side chain (or R group). In α-amino acids, the amino group is attached to the α-carbon.

- Amino acids are linked together by peptide (amide) bonds, which have double-bond character and impact the overall structure with its rigidity.

- Amino acid *residues* in proteins

- Polypeptides: peptides (up to 50 amino acids), proteins (longer)

SUGARS AND CARBOHYDRATES: MONOSACCHARIDES, POLYSACCHARIDES

- Sugars contain alcohol groups and either aldehydes (in the aldoses) or ketones (in the ketoses).

- Names may also indicate number of carbons, such as "aldohexose."

- Polysaccharides: starch and cellulose (plants), glycogen (animals)

- Nucleotides contain either ribose or deoxyribose

- Glycoproteins and glycolipids are proteins and lipids that contain carbohydrates

FATTY ACIDS AND LIPIDS

- Fatty acids contain one carboxylic acid with a long hydrocarbon chain (and usually have an even number of carbons with no branching).

- In their hydrocarbon chain, unsaturated fatty acids have at least one double bond, and saturated fatty acids have only single bonds (they're saturated with hydrogens).

- Lipids are not soluble in water.

- Triacylglycerols (three fatty acids and glycerol), used for energy storage

- Phosphoglycerides (two fatty acids, a glycerol, a phosphate, and an additional polar compound), used in cell membranes

NUCLEOTIDES AND NUCLEIC ACIDS

- Contains a five-carbon sugar (ribose or deoxyribose), a nitrogenous base, and one or more phosphate groups

- Purine bases: adenine and guanine

- Pyrimidine bases: thymine, cytosine (DNA only), and uracil (RNA only)

- ATP is a nucleotide

- Examples of *complementary base-pairing*: bases in DNA hydrogen-bond to bases in free ribonucleoside triphosphate molecules; bases in tRNA hydrogen-bond to bases in mRNA. (This is possible because of the shape and chemical properties of the bases in the nucleotide residues.)

- DNA: A=T, C≡G; RNA: A=T, G≡U

- Nucleic acids DNA and RNA contain phosphodiester linkages between nucleotides.

EXAMPLES OF EACH CLASS OF SMALL BIOMOLECULE

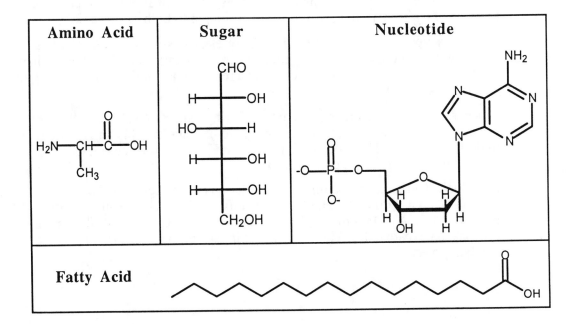

BIOCHEMICAL PROCESSES

METABOLISM = ALL ENZYME-CATALYZED REACTIONS OF A LIVING CELL (OR ORGANISM)

PRIMARY FUNCTIONS OF METABOLISM:

1. to acquire and use energy
2. to synthesize molecules needed for cellular structure and function
3. for growth and development
4. to remove waste and other toxins

Biochemical Reactions – catalyzed by enzymes

MOST COMMON REACTION TYPES AND SOME EXAMPLES:

1. Nucleophilic substitution (hydrolysis of peptides to form amino acids),
2. Elimination (removal of water - an H and an OH - to form an alkene)
3. Addition (hydration, or adding water to an alkene),
4. Isomerization (converting glucose-6-phosphate to fructose-6-phosphate (both have the same molecular formula but a different arrangement of atoms),
5. Oxidation-reduction or redox reactions (oxidizing an alcohol to an aldehyde)

BIOCHEMICAL OXIDATION AND REDUCTION: REMEMBER THE ORGANIC DEFINITIONS

In general chemistry, you learned that oxidation is the removal of electrons and that reduction adds electrons (resulting in a lower, or 'reduced,' charge or oxidation number). In organic, you learned that oxidation is the addition of oxygen (and/or the removal of hydrogen) and that reduction is the removal of oxygen (and/or the addition of hydrogen). In biochemistry, we'll tap into both of these definitions.

Redox reactions always involve a *transfer* of electrons, so when anything is oxidized, something else must be reduced. An oxidizing agent does the oxidizing (and becomes reduced in the process) and a reducing agent does the reducing (and becomes oxidized).

Biochemical redox reactions involve transferring one or two electrons at a time. Usually an H^+ rides along, so what appears to be transferred is an H atom (H•) or a hydride ion (H:⁻). (Think of the H atom as an H^+ with one electron, and the hydride ion as an H^+ with two electrons.)

Let's look at an essential example: the reduction of the coenzyme NAD^+ to produce NADH. With its additional H, we can identify NADH as the reduced form. With its extra + charge, NAD^+ is the oxidized form. When NAD^+ is reduced, something else has to be oxidized. NAD^+ is the oxidizing agent, because NAD^+ oxidizes another molecule.

Here's a specific example: NAD^+ oxidizes ethanol to form acetaldehyde:

$$NAD^+ + CH_3CH_2OH \longrightarrow NADH + H_3C-\overset{\displaystyle O}{\underset{\displaystyle H}{C}} + H^+$$

oxidized   reduced   reduced   oxidized
form   form   form   form

We could also say that ethanol reduces NAD^+ to form NADH. No matter how we look at it, two electrons (and an H^+) were transferred from ethanol to NAD^+.

Energy

Cells obtain energy by oxidizing biomolecules (or certain minerals, or by photosynthesis). Energy is in the form of electrons. *The more reduced the molecule, the more energy it contains.*

Organisms can be classified by how they obtain energy: autotrophs (photoautotrophs or chemoautotrophs), heterotrophs, chemoheterotrophs, photoheterotrophs

Metabolism = ALL ENZYME-CATALYZED REACTIONS OF A LIVING CELL (OR ORGANISM)

ANABOLIC (SYNTHESIS) VS. CATABOLIC (DEGRADATION) PATHWAYS

Anabolism is the synthesis of biomolecules. Catabolism is the breaking down of biomolecules. (*Anna* builds biomolecules, and the *cat* claws them apart.)

The removal of electrons from a molecule is *oxidation*. So, molecules are oxidized during catabolism (and reduced during anabolism). For example, consider the combustion of glucose: $C_6H_{12}O_6 + 6\,O_2 \rightarrow 6\,CO_2 + 6\,H_2O$.

The carbons in glucose ($C_6H_{12}O_6$) lose electrons to O_2 and are oxidized; the O_2 receives the electrons and is reduced.

THE CONNECTION BETWEEN CATABOLISM, OXIDATION, AND ENERGY STORAGE

When electrons are transferred during catabolism, energy is released. Cells capture this energy by driving the reaction $ADP + P_i \rightarrow ATP$. During the overall process of the oxidation of glucose, for example, the electrons are captured and put to work to make ATP. ATP supplies the energy needed for a living organism to synthesize biomolecules, to move, to maintain order, and for other functions that require energy.

The electrons are captured by redox coenzymes such as NAD^+, $NADP^+$, or FAD. For example, NAD^+ accepts two electrons (and an H^+) to form NADH. NADH transfers electrons to the *electron transport chain*, a complex system of enzymes housed along a membrane. The electrons are eventually transferred to oxygen, and the energy generated is used to pump H^+ ions to the other side of the membrane. The H^+ ions naturally flow back in through a special enzyme channel in the membrane, and as they do, the enzyme releases ATP.

Biological order is accomplished by:

1. synthesis of biomolecules
2. transport of ions and molecules across cell membranes (active transport requires energy provided by ATP hydrolysis)
3. production of motion
4. removal of metabolic waste products and other toxic substances

Living cells need a constant flow of energy to maintain this high degree of order.

OVERVIEW OF GENETIC INFORMATION PROCESSING

Central Dogma: Genetic information flows from DNA to RNA to protein.
(Retroviruses are an exception. They synthesize DNA from RNA.)

DNA directs the assembly of polypeptides from amino acids.

Gene expression = how living organisms regulate the flow of genetic information.
Genes are switched on and off so that cells can conserve resources, respond to the environment, and/or develop properly.

codon = the three-base code that translates to a specific amino acid (For example, the base sequence uracil-guanine-uracil codes for the amino acid cysteine.)

In this phase of gene expression:	*and from the base sequence information in:*	*this is synthesized:*	*using the primary tool(s):*
Transcription	DNA genes	RNA	RNA polymerase
Translation	mRNA	polypeptides	ribosomes (rRNA, ribosomal proteins) tRNA

HOW DOES TRANSLATION WORK?

1. The tRNA anticodon complementary base-pairs with the mRNA codon. The tRNA is carrying an amino acid.
2. The amino acid held by the tRNA forms a peptide bond with the previous amino acid of the growing protein.
3. The tRNA releases its amino acid.
4. Translocation: The mRNA moves relative to the ribosome, and a new codon enters the catalytic site.
5. This repeats until a termination or stop codon signals the release of the polypeptide.

BIOCHEMICAL METHODS 1.1: AN INTRODUCTION

Reductionism: to understand a complex system, analyze its simplest components.

ORGANIC CHEMISTRY CONNECTION

Functional groups that may be new to you

R–SH   Thiol group, also known as a sulfhydryl group

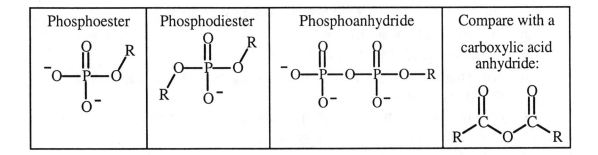

Phosphoester	Phosphodiester	Phosphoanhydride	Compare with a carboxylic acid anhydride:

AFTER STUDYING THIS CHAPTER, YOU SHOULD BE ABLE TO SOLVE THESE TYPES OF PROBLEMS:

- Identify functional groups in a given molecule.

- Recognize the structures of the four classes of biomolecules. Classify a given compound as an amino acid, sugar, fatty acid, or nucleotide.

- Recognize the four types of reactions: nucleophilic substitution, elimination, oxidation-reduction, and addition.

CHAPTER 1: ANSWERS TO EVEN-NUMBERED REVIEW QUESTIONS

2. The three domains of living organisms are the Archaea, the Bacteria and the. Eukarya. The Archaea are best known for the capacity of many of their species to thrive in extremely hostile environments. Bacterial species are characterized by their vast biochemical diversity. The Eukarya are extraordinarily complex organisms, many of which are multicellular.

4. The functional group(s) in each molecule are

    a. aldehyde group

    b. carboxylic acid and amino groups

    c. sulfhydryl group

    d. ester group

    e. alkene

    f. amide group

    g. ketone group

    h. alcohol group

6. a. biochemistry – the study of the molecular basis of life

    b. oxidation – loss of an electron

    c. reduction – gain of an electron

    d. active transport – movement of substances against a concentration gradient; energy is required

    e. leaving group – a molecular group displaced during a reaction

    f. elimination – loss of an atom or group

    g. isomerization – a shift of atoms or groups within a molecule

    h. nucleophilic substitution – displacement of an atom or group by an electron-rich species

    i. reducing agent – atom or group oxidized during an oxidation/reduction reaction

    j. oxidizing agent – atom or group reduced during an oxidation/reduction reaction

8. DNA is the repository of each organism's genetic information. RNA is the nucleic acid that is involved in the expression of genetic information, primarily in various aspects of protein synthesis.

10. Plants dispose of waste products either by degradation or storage in vacuoles or cell walls.

12. The most important advantages of multicellular organisms over unicellular organisms are: multicellular organisms have a stable internal environment; cells can be specialized (division of labor); there is a more efficient use of resources; sophisticated functions can be accomplished by such organisms.

14. a. metabolism – the sum of the chemical reactions carried out in a living cell

    b. nucleophile – an atom or group with an unshared pair of electrons that is involved in a displacement reaction

    c. reductionism – complex processes can be understood by examining their simpler parts

    d. electrophile – an electron-deficient species

    e. energy – the ability to do work

16. The primary functions of metabolism are acquisition and utilization of energy, synthesis of biomolecules, and removal of waste products.

18. Important ions found in living organisms are Na^+, K^+, Ca^{2+}, and Cl^-.

20. The functions of polypeptides include transport, structural composition, and catalysis (enzymes).

22. The largest molecules in living organisms are the nucleic acids and the proteins. Nucleic acids store genetic information (DNA) and mediate the synthesis of proteins (RNA). The proteins are the tools which perform all of the tasks required to sustain living processes.

24. Order is maintained in living organisms primarily by the synthesis of biomolecules, the transport of ions and molecules across cell membranes, the production of force and movement, and the removal of metabolic waste products.

26. In a hierarchical system, such as that found in living organisms, emergent properties (in any level in the structural hierarchy) have characteristics that cannot be predicted from the analysis of its component parts. One example is the structural and functional properties of a protein such as hemoglobin, which is composed of carbon, nitrogen, hydrogen, oxygen, and iron atoms.

CHAPTER 1: ANSWERS TO EVEN-NUMBERED THOUGHT QUESTIONS

2. The assumption that the biochemical processes in prokaryotes and eukaryotes are similar is only safe when basic living processes are considered (e.g., glycolysis and the general principles of genetic inheritance). Living organisms are so diverse in their adaptations that information acquired from research with prokaryotes must be judiciously interpreted in reference to eukaryotes.

4. The amide carbon-nitrogen bond has double bond character. This restricts bond rotation about the bond, making the bond much more rigid and the protein less flexible.

    This can be shown by drawing the resonance structure for an amide bond. Although this resonance structure may seem unlikely because of the charge separation, it is indeed significant, and explains the observed rigidity of the C–N bond quite well.

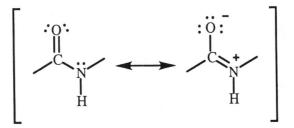

6.   Micelle: Bilayer:

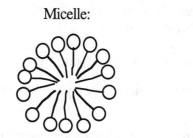 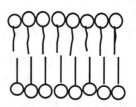

    The structure of a micelle is illustrated in Figure 3.11 (p. 74) in your text. The lipid bilayer is a prominent feature of Figure 2.2 (p. 32) in your text.

8.   The capacity of healthy bodies to adapt to high cholesterol diets by inhibiting cholesterol synthesis is an example of the capacity of living organisms to regulate their metabolic processes.

CHAPTER TWO

Living Cells

This is where it all happens. With a clear picture of a living cell, many of the structures and functions of specific biomolecules and their metabolic pathways will make much more sense. Understanding the composition of cells and their organelles gives meaning to the organization of life on a molecular scale.

Don't underestimate the power of creating your own diagrams and your own tables. Organize the material in a way that will help you the most, whether as pictures or diagrams of the organelles and cells, as lists that include functions, as summaries, or as tables.

Question 2.4, on page 57, is a terrific study tool. Also, try a post-it note technique: write the name of an organelle on one side of a post-it note, and its function on the reverse. Notes of different colors can help keep prokaryotic- or plant- specific features separate.

CREATE A TABLE in which each row is devoted to a specific organelle. Check out the following example. You'll want to tailor the column headers to the emphasis given in your specific class or to challenges that you're facing in learning more about these great little packages.

Item:	Structure:	Location:	Function:	Notes:
What is it?	What's it made of?	Where is it?	What does it do?	Why is it important? How is it unique?

Suggested Item Column: Plasma Membrane, Nucleus, RER, Ribosomes, Golgi Apparatus, Lysosomes, Peroxisomes, Mitochondria, Plastids

---

BASIC THEMES

Water

Hydrophilic vs. hydrophobic portions of molecules affect their overall structure. Example: Proteins fold so that hydrophobic areas hide inside and the hydrophilic areas face outward (towards the aqueous environment).

Biological Membranes as selective physical barriers

Importance of the control of transport across membranes

Basic structure: lipid bilayer (mostly phospholipids) with proteins

Phospholipids: hydrophilic head, hydrophobic tails

Membrane protein functions include: transport of molecules and ions, energy generation, signal transduction

Self-Assembly into supramolecular structures

Based on hydrophilic and hydrophobic interactions; large numbers of weak interactions

Molecular chaperones or templates may provide assistance with assembly (folding)

Molecular Machines

Proteins (and protein complexes) that perform work, consist of moving parts, require energy-transducing mechanisms (convert energy into directed motion)

Nucleotide (e.g. ATP or GTP) binds to motor proteins (protein subunits); nucleotide hydrolyzes and releases energy, causing precisely targeted change in the protein subunit's 3-D shape; change is transmitted to nearby subunits

Motor proteins

Examples of molecular machines:

ribosomes - rapidly and accurately incorporate amino acids into polypeptides

sarcomeres - contractile units of skeletal muscle; actin, myosin are proteins

---

STRUCTURE OF PROKARYOTIC CELLS

Features: relatively small, able to move using pili or flagella, able to retain specific dyes

Identify based upon: nutritional requirements, energy sources, chemical composition, biochemical capacities

Common Features: Cell Wall   Circular DNA molecules

Plasma Membrane   No internal membrane-enclosed organelles

Cell wall

Glycocalyx (slime layer or capsule), peptidoglycan

Gram positive cells have a thick peptidoglycan layer

Plasma membrane

Inner membrane / periplasmic space / outer membrane

May contain proteins for photosynthesis and respiration

Cytoplasm

Nucleoid - contains a chromosome (circular DNA molecule)

Plasmids - additional small circular DNA molecules

Ribosomes, Inclusion bodies

Pili and Flagella

motion and conjugation (singular: pilus and flagellum)

STRUCTURE OF EUKARYOTIC CELLS

Plasma membrane

Controls transport of molecules into and out of the cell

Transport is facilitated by carrier and channel proteins

Glycocalyx, Receptors, Extracellular matrix

Nucleus

Contains cell's hereditary information, regulates metabolism by directing the synthesis of protein cell components

Nucleoplasm; lamins, chromatin fibers, histones

Nuclear envelope; nuclear pore; nuclear pore plug

Nucleolus: synthesis of ribosomal RNA

Endoplasmic Reticulum (ER)

Lumen (or cisternal space)

Rough ER (RER): synthesis of membrane proteins and proteins for export from the cell; contains ribosomes

Smooth ER (SER): lipid synthesis, biotransformation

Ribosomes: protein synthesis

Golgi Apparatus (or Golgi complex)

Packages and distributes cell products to internal and external compartments

Vesicles: secretory vesicles (or secretory granules)

Cisterna (plate); cis and trans faces

Exocytosis (secretion, figure 2.20)

Lysosomes: intracellular and extracellular digestion

Contain digestive enzymes (acid hydrolases)

Endocytosis

Endosome (lysosome precursor)

Lysosomal storage diseases (e.g. Tay-Sachs disease)

Peroxisomes: contain oxidative enzymes

Most important function: generate and break down peroxides (R-O-O-R)

Mitochondria:

Aerobic respiration - oxygen-dependent synthesis of ATP

Outer membrane, inner membrane, cristae (folds); intermembrane space; matrix

Respiratory assemblies: ATP synthesis

Plants only:

Vacuole (contains acid hydrolases, analogous to lysosomes in animals)

Microfilaments

Cell wall (contains cellulose)

Dictyosomes (analogous to Golgi apparatus in animal cells)

Peroxisomes: two types: 1. in leaves; responsible for photorespiration, and

2. glyoxysomes, in germinating seed; convert lipids to carbohydrates

Plastids are analogous to mitochondria in animal cells

plants, algae, some protists

proplastids (plastid precursors)

plastid types: 1. leucoplasts (storage);

2. chromoplasts (accumulate plant pigments)

chloroplasts: photosynthesis (conversion of light energy into chemical energy)

thylakoid membrane, grana, thylakoid lumen (or channel)

stroma (analogous to mitochondrial matrix), stroma lamellae

Cytoskeleton

Functions: 1. maintains cell shape, 2. facilitates coherent cellular movement,

3. provides supporting structure to guide organelle movement within the cell

Microtubules: structural support for long, thin cells; protein = tubulin

Microfilaments: cytoplasmic streaming and ameboid movement; protein = actin

Intermediate fibers: maintain cell shape under mechanical stress; various proteins
(Example: keratin filaments in outer skin cells)

14

SPECIAL INTEREST BOXES:

Endosymbiosis: symbiosis, anaerobic, aerobic respiration

The Origin of Life: abiogenesis, RNA world concept

CHAPTER 2: ANSWERS TO EVEN-NUMBERED REVIEW QUESTIONS

2. All living cells have similar chemical compositions (i.e., they are all composed of molecules such as carbohydrates, proteins, and lipids) and they all utilize DNA as genetic material.

4. The plasma membranes of both prokaryotic and eukaryotic cells control the flow of substances into and out of the cell. In addition, plasma membrane receptors bind to specific molecules in the cell's external environment. In prokaryotes, for example, some receptors allow the organism to respond to the presence of food molecules. In eukaryotes, numerous cell receptors bind specific hormone or growth factor molecules. The prokaryotic cell wall is sufficiently rigid that it maintains the organism's shape and protects against mechanical injury.

6. a. Exocytosis - a cellular process which consists of the fusion of membrane-bound secretary organelles with the plasma membrane. The contents of the granules are then released into the extracellular space.

    b. Biotransformation - a biochemical process in which water-insoluble organic molecules are prepared for excretion.

    c. Grana - tightly stacked portions of thylakoid membrane within chloroplasts.

    d. Symbiosis - the living together of two dissimilar organisms in an intimate relationship.

    e. Self-assembly - the formation of supermolecular complexes made possible by the interactions of specific biomolecules each of which has an intricately shaped surface

    f. Hydrophobic - refers to molecules that possess few, if any, electronegative changes; do not dissolve in water

    g. Hydrophilic – refers to molecules that possess positive or negative charges or contain relatively large numbers of electronegative oxygen or nitrogen atoms; dissolves easily in water

    h. Motor protein - a component of a biological machine that binds nucleotides; nucleotide hydrolysis drives precise changes in the protein's shape

    i. Endosymbiosis - a mechanism that has been proposed to explain the evolution of modern eukaryotic cells. In the hypothesis, primordial eukaryotic cells engulfed smaller prokaryotic cells that eventually became modern mitochondria, chloroplasts and possibly other cell structures such as flagella and cilia.

    j. Proplastids - small nearly colorless plant cell structures that develop into the plastids of differentiated cells.

    k. Thylakoid - an intricately folded membrane system that is responsible for several chloroplast metabolic functions.

8. Plant cells; leucoplasts; chromoplasts.

10. The highly developed framework of the cytoskeleton performs the following functions in eukaryotic cells: (1) maintenance of overall cell shape, (2) facilitation of coherent cellular movement, and (3) provision of a supporting structure that guides the movement of organelles within the cell.

12. The nucleus is the repository of the cell's hereditary information. The nucleus also exerts a profound influence over all the cell's metabolic activities.

14. The principal function of the rough endoplasmic reticulum is the synthesis of membrane proteins and protein for export from the cell. Smooth endoplasmic reticulum, so-named because it lacks attached ribosomes, is involved in lipid synthesis and biotransformation processes.

CHAPTER 2: ANSWERS TO EVEN-NUMBERED THOUGHT QUESTIONS

2. Specialized cells can perform very sophisticated functions that make multicellular organisms possible. Cell specialization can be considered a disadvantage because such cells cannot exist independently; that is, they can only exist as part of a multicellular organism where their metabolic needs (e.g., energy requirements and waste product removal) are met.

4. The immobilization of enzymes and organelles on the cytoskeleton facilitates the highly organized set of living processes required to sustain the living state. For example, the close proximity of immobilized enzymes in a biochemical pathway allows the rapid delivery of the product of one enzyme to the active site of the next. This circumstance requires lower concentrations of reactant molecules than the time consuming diffusion process.

6. The volume of a spherical mycoplasma cell is calculated with the equation $V = 4\pi r^3/3$. Since the diameter $= 0.3$ μm, the radius $= 0.15$ μm.

   $V = 4\pi r^3/3 = (4)(3.14)(0.15)^3/3$

   $= 4(3.14)(0.00337)/3$

   $V = 0.014$ μm^3

   Assuming that *E. coli* is cylindrical with dimensions of 1 μm x 2 μm (i.e., a typical rod-shaped bacteria), its volume would be $\pi r^2 h = (3.14)(0.5)^2(2) = 1.6$ μm^3 (Refer to the answer to in-chapter question 2.1.)

   At 0.014 μm^3, the volume of a typical mycoplasma is significantly smaller than the volume of the cylindrical *E.coli* cell. (To be more specific, the mycoplasma is 0.9 % of the size of an E. coli. Alternatively, the E. coli is about 114 times larger than the mycoplasma.)

CHAPTER THREE

Water: The Medium of Life

Without the unique properties of water, life as we know it could not exist. This chapter serves as a refresher for a number of concepts that you've seen in general and organic chemistry (electronegativity, polarity, hydrogen bonds, heat capacity, osmotic pressure, acid-base chemistry, buffers) and places them solidly in the context of biological systems. If some of these concepts happened to slip by you in previous courses, chances are that you'll find them more interesting and accessible here. You'll be seeing much of this material applied to amino acids, proteins, enzymes, and in the chapters beyond, so be sure to practice (and learn) this material early. This chapter includes examples of buffer problems with detailed solutions, and extra practice problems.

MOLECULAR STRUCTURE OF WATER - H_2O

- Electronegativity, polar bonds (C–O, C–N, H–O), polar molecules, dipoles

  Whether or not a molecule is polar also depends on its geometry (or overall shape). For example, H_2O is polar because it's bent at an angle of 104.5°, but CO_2 is nonpolar. The polarity of the two oxygen–carbon bonds cancels out because CO_2 is linear: O=C=O.

- The polar nature of water allows it to interact with a variety of other molecules. For example, table salt (NaCl) dissolves completely in water because of ion-dipole interactions. Alcohol dissolves in water via dipole-dipole interactions.

- electrostatic interactions

NONCOVALENT BONDING

Ionic Interactions, salt bridges

Hydrogen Bonds

- Occur between a hydrogen (that is attached to an oxygen or nitrogen) and a lone pair of electrons

- Hydrogen bonding of water explains: relatively high melting and boiling points, heat of vaporization, heat capacity, surface tension, and viscosity

van der Waals Forces

- Dipole – dipole interactions, dipole – induced dipole interactions, induced dipole – induced dipole interactions (London dispersion forces)

- The easier an atom can be polarized, the stronger the van der Waals forces.

THERMAL PROPERTIES OF WATER

- Higher-than-expected melting and boiling points due to hydrogen bonding

- High heat of vaporization - water doesn't boil easily

- High heat capacity - water can absorb and store heat and release it slowly

- Thermal regulation of living organisms: high water content (helps to retain heat due to high heat capacity), evaporation as a cooling mechanism

SOLVENT PROPERTIES OF WATER

Hydrophilic molecules (water-loving)

- Polar and ionic molecules

- Solvation spheres (formed by water around solutes)

Hydrophobic molecules (water-hating)

- Nonpolar molecules

- Hydrophobic effect: Hydrophobic interactions result from the energetics of the surrounding water molecules. Water prefers to form hydrogen bonds with other water molecules, excluding the hydrophobic molecules.[1]

Amphipathic molecules (contain both hydrophobic and hydrophilic ends)

- Molecules with a hydrophilic "head" and a long hydrophobic "tail" form micelles or bilayers (example: phospholipids form bilayers to create biological membranes)

Osmotic pressure (π)

- Osmosis, osmometer

- $\pi = i\mathrm{MRT}$,   where i = degree of ionization (van't Hoff factor)

        M = molarity, R = 0.082 L·atm/K·mol, T = temperature (Kelvin)

- Osmolarity = iM

- Isotonic vs. hypertonic vs. hypotonic; hemolysis vs. crenation

- Membrane potential and the Donnan effect

- Cells regulate their osmolarity (by pumping ions or by synthesizing osmolytes)

- Calculation of molecular weight using osmotic pressure data

---

[1] Also, in Chapter 4 you'll learn that the overall disorder of the water increases when a lipid micelle or membrane is formed (or when a protein folds), and that the water's disorder outweighs the resulting increased order of the micelle.

IONIZATION OF WATER

Acids, Bases, and pH

$$HA \rightleftharpoons H^+ + A^-$$

- Strong vs. weak acids and bases

- weak acid (HA) and conjugate base (A^-)

- K_a is the equilibrium constant for the loss of an H^+

$$K_a = \frac{[H^+][A^-]}{[HA]}$$

- pK_a: the lower the pK_a, the stronger the acid (see below)

The stronger the acid, the more dissociated the acid is. That means there's more H^+ in solution, so the K_a will be larger. But remember that the $pK_a = -\log K_a$. If the K_a is 10^{-4}, the pK_a is 4. For a K_a that's a hundred times greater at 10^{-2}, the pK_a is 2. So, the lower the pK_a, the stronger the acid.

Buffers

- Buffer = solution of a weak acid (HA) and its conjugate base (A^-)

- Buffers resist pH changes when H^+ or OH^- is added (Le Chatelier's principle)

- Acidosis vs. alkalosis

BUFFERING CAPACITY

depends on total buffer concentration and ratio of $[A^-]/[HA]$

(total buffer concentration = $[HA] + [A^-]$)

HENDERSON-HASSELBALCH EQUATION: $pH = pK_a + \log \dfrac{[A^-]}{[HA]}$

Buffers are most effective when $[A^-] = [HA]$ or in the pH range of $pK_a \pm 1$. Titration curves show relatively flat areas when the $pH = pK_a$. These flat areas indicate good buffer ranges because a relatively large amount of OH– may be added with very little change in pH.

WEAK ACIDS WITH MORE THAN ONE IONIZABLE GROUP

Examples: phosphoric acid, amino acids

PHYSIOLOGICAL BUFFERS

BICARBONATE BUFFER

The bicarbonate buffer system is slightly more complicated that one would expect because CO_2 reacts with H_2O to form H_2CO_3.

$$CO_2 + H_2O \rightleftharpoons H_2CO_3 \rightleftharpoons H^+ + HCO_3^-$$

This can be simplified to:

$$CO_2 + H_2O \rightleftharpoons H^+ + HCO_3^- \qquad\qquad pK_a = 6.37$$

Given that its pK_a differs from blood pH (7.4) by more than one pH unit, how can bicarbonate serve as an important buffer in the blood?

Both CO_2 and HCO_3^- can be regulated. CO_2 can be exhaled, and the kidneys can remove H^+ or HCO_3^- from the blood as needed.

PHOSPHATE BUFFER

$$H_2PO_4^- \rightleftharpoons H^+ + HPO_4^{2-} \qquad\qquad pK_a = 7.2$$

Given that the pK_a of dihydrogen phosphate is so close to blood pH (7.4), why *isn't* $H_2PO_4^-/HPO_4^{2-}$ an important buffer system in the blood?

Their concentrations are too low. Phosphate buffers *are* important in intracellular fluids, where $[H_2PO_4^-]$ and $[HPO_4^{2-}]$ are higher.

PROTEIN BUFFERS

Some amino acid side groups are weak acids or bases, so some proteins can act as buffers. For example, hemoglobin, serum albumins, and other proteins act as buffers to help regulate the pH of the blood.

SOLVING BUFFER PROBLEMS: USE THE HENDERSON-HASSELBALCH EQUATION

$$pH = pK_a + \log\frac{[A^-]}{[HA]}$$

$$\frac{[A^-]}{[HA]} = \frac{\text{conjugate base}}{\text{weak acid}}$$

Total buffer concentration = $[A^-] + [HA]$

One way to remember this equation is:

H comes before K, and AHA!!

pH is equal to pK_a plus the log of AHA!!

(Yell the Aha! with enthusiasm and you'll always remember this equation!)

A buffer is a mixture of a weak acid and its conjugate base. The toughest part of solving buffer problems is often identifying the weak acid (HA) and its conjugate base (A^-). Remember that an acid donates an H^+ and a base accepts the H^+. So, the weak acid will always have an extra H^+ when compared to its conjugate base, and the conjugate base will have *one* extra negative charge.

Examples:

__HA (weak acid)__	__A^- (conjugate base)__
H_2CO_3	HCO_3^-
$H_2PO_4^-$	HPO_4^{2-}

These examples were chosen for two reasons: 1) they are present in living systems as physiological buffers, and 2) they can be confusing because one of the forms listed in each example above can serve as either an acid or a conjugate base, depending what else is present (and the pH). Pay close attention to the problem, and be sure that you identify the acid as the one with the extra H^+. For example, HCO_3^- is the conjugate base above, but if CO_3^{2-} was present, then HCO_3^- would be the weak acid.

__HA (weak acid)__	__A^- (conjugate base)__
HCO_3^-	CO_3^{2-}
H_3PO_4	$H_2PO_4^-$

Hint: It helps to write out the acid-base reaction in this form. Be sure that both the number of H's and the charges balance.	$HA \rightleftharpoons H^+ + A^-$
Check out these examples. Writing reactions like this helps to make it more clear which is HA and which is A^-. Note that $H_2PO_4^-$ is the conjugate base in the first equation, but it's the weak acid in the second equation.	$H_3PO_4 \rightleftharpoons H^+ + H_2PO_4^-$ $H_2PO_4^- \rightleftharpoons H^+ + HPO_4^{2-}$
This carboxylic acid group is the weak acid, and its carboxylate is the conjugate base.	$R-\overset{\displaystyle O}{\overset{\|}{C}}-OH \rightleftharpoons H^+ + R-\overset{\displaystyle O}{\overset{\|}{C}}-O^-$
An amine group is a conjugate base, and its conjugate acid is shown here on the left.	$R-NH_3^+ \rightleftharpoons H^+ + R-NH_2$

HOW TO PREPARE A BUFFER GIVEN A TARGET pH AND TOTAL CONCENTRATION

Use the Henderson-Hasselbalch equation to solve for [A⁻]/[HA]. Rearrange your answer in the form: [HA] = x [A⁻] With the equation for the total concentration of a buffer, you can solve for [A⁻] by substituting "x [A⁻]" for the value of [HA]. Once you have [A⁻], you can use either equation to solve for [HA]. See example problem # 3, below.

Also, note that you can't buy "A⁻." It typically comes as a sodium or potassium salt (NaA or KA). For example, the buffer system HPO_4^{2-}/ $H_2PO_4^-$ would be prepared by dissolving the appropriate amount of NaH_2PO_4 and Na_2HPO_4 in water, then diluting to the final volume. So, the final step often involves converting moles to grams of salt, using the molecular weight.

WHAT HAPPENS WHEN YOU ADD A STRONG ACID TO A CONJUGATE BASE...

...either in a buffer solution or alone? Write out the acid-base equation and think about what's happening. [Hint: Be sure to work in moles (or millimoles).] For example, think about what would happen if you added 0.02 moles of strong acid (H^+) to 0.05 moles of conjugate base (A^-). The 0.02 moles of H^+ would react completely with 0.02 moles of A^-, forming).02 moles of HA, and leaving (0.05 – 0.02) = 0.03 moles of A^- left over.

EXAMPLES OF BUFFER PROBLEMS *(DETAILED SOLUTIONS FOLLOW)*

1. What is the pH of a phosphate buffer when [$H_2PO_4^-$] = 20 mM and [HPO_4^{2-}] = 15 mM? The pK_a = 7.20 *(from Table 3.4, page 80 in your text)*.

2. Calculate the ratio of $\dfrac{[HPO_4^{2-}]}{[H_2PO_4^-]}$ at pH 6.2, 7.2, and 8.2. The pK_a is 7.2.

3. Describe how you would prepare one liter of 0.20 M lactate buffer with a pH of 4.2. What ratio of lactate salt to lactic acid would you use? The pK_a of lactic acid is 3.86.

4. a. What is the pH of a solution prepared by mixing 100 mL of 1.00 M HCl with 300 mL of 0.500 M sodium succinate? The pK_{a1} of succinic acid is 4.21.

    b. Is this buffer system effective? If so, why? If not, how could you correct it?

    c. Oops! Somebody just sploshed 80 mL of 1.0 M NaOH into your carefully-made solution! What's the final pH? Did the buffer work? How do you know? Is the solution still a good buffer?

EXAMPLES OF BUFFER PROBLEMS: *SOLUTIONS*

1. *What is the pH of a phosphate buffer when [$H_2PO_4^-$] = 20 mM and [HPO_4^{2-}] = 15 mM? The pK_a = 7.20.*

    First, identify the acid and the conjugate base: $H_2PO_4^- \rightarrow H^+ + HPO_4^{2-}$
    Since $H_2PO_4^-$ donates an H^+, it is the acid and HPO_4^{2-} is the conjugate base.

    Use the Henderson-Hasselbalch equation to determine the pH of the buffer solution.

    $$pH = pK_a + \log \frac{[HPO_4^{2-}]}{[H_2PO_4^-]} = 7.20 + \log 0.75 = 7.20 - 0.12 = 7.08$$

    Always think about your answer to be sure that it makes sense. Compare the amount of acid vs. base present with the pH vs. pK_a. For example, in this problem we have more acid than base, and at 7.08, the final pH is more acidic than the pK_a. Yay!

2. *Calculate the ratio of* $\frac{[HPO_4^{2-}]}{[H_2PO_4^-]}$ *at pH 6.2, 7.2, and 8.2. The pK_a is 7.2.*

    Identify the acid and the conjugate base: $H_2PO_4^- \rightleftharpoons HPO_4^{2-} + H^+$

    As before, $H_2PO_4^-$ is the acid and HPO_4^{2-} is the conjugate base. Plug the pH and pK_a values into the Henderson-Hasselbalch equation and solve for the ratio $\frac{[A^-]}{[HA]}$.

    $$pH = pK_a + \log \frac{[A^-]}{[HA]}$$

    At pH 6.2: $\log \dfrac{[A^-]}{[HA]} = pH - pK_a = 6.2 - 7.2$

    $$\frac{[A^-]}{[HA]} = 10^{(6.2-7.2)} = 10^{-1} = 0.1 \text{ or } \frac{1}{10}$$

    At pH 7.2: $\dfrac{[A^-]}{[HA]} = 10^{(7.2-7.2)} = 10^0 = 1 \text{ or } \dfrac{1}{1}$

    At pH 8.2 $\dfrac{[A^-]}{[HA]} = 10^{(8.2-7.2)} = 10^1 = 10 \text{ or } \dfrac{10}{1}$

3. *Describe how you would prepare a 0.20 M lactate buffer with a pH of 4.2. What ratio of lactate salt to lactic acid would you use? The pK_a of lactic acid is 3.86.*

Use the Henderson-Hasselbalch equation to calculate the ratios of the salt and acid, where the salt is "A⁻" and the acid is "HA".

$$pH = pK_a + \log \frac{[A^-]}{[HA]}$$

Substituting these values into the equation gives:

$$4.2 = 3.86 + \log [A^-]/[HA]$$
$$0.34 = \log [A^-]/[HA]$$
$$10^{0.34} = 2.19 = [A^-]/[HA]$$
$$[A^-] = (2.19)[HA] \qquad \text{(equation 1)}$$

Since the total concentration of the lactate buffer is 0.20 M, we know that:

$$[A^-] + [HA] = 0.2 \text{ M} \qquad \text{(equation 2)}$$

Using simultaneous equations (i.e., substituting the value for [A⁻] from equation 1 into equation 2) to determine the concentrations of the salt and acid for this particular buffer solution gives:

$$(2.19)[HA] + [HA] = 0.20 \text{ M}$$
$$(3.19)[HA] = 0.20 \text{ M}$$
$$[HA] = 0.063 \text{ M}$$

Take this value and inserting it into (equation 1) to solve for [A⁻]:

$$[A^-] = (2.19)[HA] = (2.19)(0.063) = 0.14 \text{ M}$$

To prepare this buffer, place 0.14 moles of lactate (or sodium lactate) and 0.063 moles of lactic acid in a 1 L volumetric flask and dilute with water to the 1 L mark.

And now to double-check: We're adding more base than acid to make this buffer. Since our final pH, 4.2, is more basic than our pK_a, our answer makes sense. Yay!

4. *a.* *What is the pH of a solution prepared by mixing 100 mL of 1.00 M HCl with 300 mL of 0.500 M sodium succinate? The pK_{a1} of succinic acid is 4.21.*

First, identify the weak acid and its conjugate base. Sodium succinate would give Na+ and succinate⁻ in solution. HCl is a strong acid that would react with the succinate⁻ to form H-succinate, or succinic acid. So, the weak acid would be succinic acid and the conjugate base would be succinate⁻.

The number of moles of succinate⁻ initially is (300 mL)(0.500 M) = 150 mmoles

The number of moles of HCl added is (100 mL)(1.00 M) = 100 mmoles

It's a good assumption that HCl will react completely with the succinate⁻, so that would give: 150 mmoles succinate⁻ – 100 mmoles reacted with HCl

= 50 mmol succinate⁻ left over and 100 mmol of succinic acid formed.

Now, use the Henderson-Hasselbalch equation: $pH = pK_a + \log \dfrac{[\text{succinate}^-]}{[\text{succinic acid}]}$

$$pH = 4.21 + \log \frac{[50 \text{ mmol}]}{[100 \text{ mmol}]} = 4.21 - 0.30 = \mathbf{3.91}$$

A pH of 3.91 makes sense because more acid than base is present, and 3.91 is more acidic than 4.21, the pK_a.

Note that this method included a shortcut - mmoles rather than molarity was used. This is valid because the total volume is the same and would cancel out. To use molarity, divide both the numerator (0.0045 moles) and the denominator (0.0030 moles) by the total volume, 0.450 L (300 mL+150 mL).

b. *Is this buffer system effective? If so, why? If not, how could you correct it?*

This is a good buffer because its pH is within the range of the pK_a ± one pH unit, or 3.21 to 5.21. If the pH was lower than 3.21, more succinate should be added. If the pH was higher than 5.21, more acid should be added.

c. *Oops! Someone just splooshed 80 mL of 1.0 M NaOH into your carefully-made solution! What's the final pH? Did the buffer work? How do you know? Is the solution still a good buffer?*

pH = 4.61, yes (see below), yes

From (a), we know that before the addition of NaOH, we have 50 mmol of succinate and 100 mmol of succinic acid. The OH⁻, a strong base, will react completely with the succinic acid to form succinate. We have (80 mL)(1.0 M) = 80 mmol OH⁻. So, we will have:

100 mmol succinic acid – 80 mmol reacted (with OH⁻) = 20 mmol succinic acid

50 mmol succinate + 80 mmol formed = 130 mmol succinate

Let's plug these new values into the Henderson-Hasselbalch equation:

pH = 4.21 + log(130 mmol/20 mmol) = 4.21+0.81 = **5.02** (which makes sense!)

The buffer worked. If 80 mmol of OH⁻ were present in the same volume of pure water, the final pH would be 12.5.

It's still a pretty good buffer since 5.02 is less than one pH unit from the pK_a. However, it will have a greater buffering capacity for acids than for bases.

SOAPS: EXAMPLES OF AMPHIPATHIC MOLECULES

Soaps have a long hydrocarbon chain (tail) and a carboxylic acid (head) that replaced its H⁺ with a Na⁺ or K+. Soaps are salts of fatty acids.

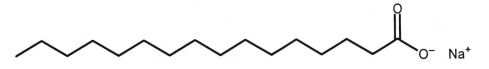

Why are soaps effective at getting rid of grease? The hydrophobic tail interacts with the non-polar "grease" molecule. This allows the polar end of the soap to interact with water. Water is then able to completely surround the grease molecule, so the grease/detergent micelle becomes water-soluble and washes away.

AFTER STUDYING THIS CHAPTER, YOU SHOULD BE ABLE TO SOLVE THESE TYPES OF PROBLEMS:

Noncovalent Interactions

- Identify the types of noncovalent interactions that can occur between given molecules (or between a given molecule and water).

- Identify whether or not a given molecule has a dipole moment.

- Will a micelle be able to form in a solution of a given molecule?

Acids and Bases

- Identify weak acid-conjugate base pairs.

- Calculate pH from a given [H$^+$], or [H$^+$] from a given pH.

Buffers: Use the Henderson-Hasselbalch equation

- Identify mixtures that can form buffer systems, and give the pH range where each would be most effective.

- Calculate the pH of a buffer given the pK_a and amounts of a weak acid and its conjugate base. (Calculate the amounts of a weak acid and its conjugate base given the amounts of conjugate base and HCl, *or* given the amounts of weak acid and NaOH.)

- Solve for the ratio of [A$^-$]/[HA].

- Determine how to prepare a specific buffer given the target pH, total buffer concentration, and total volume.

- Calculate the pH of a buffer solution after HCl or NaOH has been added.

- Titration curves: Estimate the pK_a and the effective buffer range.

Osmotic pressure:

- $\pi = i$MRT, R = 0.082 L·atm/K·mole

- Calculate osmotic pressure given the amount of solute and volume.

- Calculate osmolarity (iM)

- Given osmotic pressure data, calculate the molecular weight.

- Predict the direction that water will flow during dialysis. Predict whether a cell will shrink or swell, given various changes in osmotic pressure.

PRACTICE PROBLEMS TO PROMOTE PERFECTION

Answers are at the very end of this chapter.

1. Describe how you would prepare a 0.40 M acetate buffer with a pH of 5.4. What ratio of acetate salt to acetic acid would you use? The pK_a of acetic acid is 4.76.

2. An acetate buffer is prepared by adding 40.0 mL of 1.00 M acetic acid to 200 mL of an aqueous solution containing 2.00 grams of sodium acetate. The final solution is diluted to 300 mL. (The molecular weight of sodium acetate is 82.0 g/mole, and the pKa of acetic acid is 4.76.)

   a. What is the pH?
   b. What is the total buffer concentration?
   c. What would the final pH be if 20 mL of 1.0 M of hydrochloric acid were added to this buffer solution? Is this still a good buffer? What would the pH have been if the buffer hadn't been present? (Assume the same total volume of water.)

3. For laboratory experiments that are extremely sensitive to pH, why is it often recommended to use *freshly* distilled water?

4. An 8-oz. serving (240 mL) of a popular cola contains 27 grams of sugars, listed as "high fructose corn syrup and/or sugar." What is the osmotic pressure that would be exerted by 27 grams of fructose in 240 mL of water at 37°C? If the sugars were 27 grams of sucrose, how would that affect the osmotic pressure? The molecular weight of fructose is 180 g/mole, and the molecular weight of sucrose is 342 g/mole.[2]

5. That same 240 mL serving of cola contains 25 mg sodium. If we assume that the sodium is present as sodium chloride,[3] that corresponds to 4.53×10^{-3} M NaCl. What would be the osmolarity and osmotic pressure of an aqueous solution of NaCl at this concentration at 37°C? Assume 100% ionization of the NaCl.

6. A solution of 0.200 grams of an unknown molecule in 100 mL of water exerts an osmotic pressure of 0.465 atm at 25°C. Calculate the molecular weight of this nonelectrolyte.

---

[2] Obviously, this isn't a very good approximation for the cola, since the sugars are a mixture and there are other ingredients, such as caffeine, phosphoric acid, citric acid, carbonation, and "natural flavors."

[3] O.k., so it's probably a terrible assumption, especially since "salt" isn't listed in the ingredients.

CHAPTER 3: ANSWERS TO EVEN-NUMBERED REVIEW QUESTIONS

2. $pH = -\log [H^+]$

    $8.3 = -\log [H^+]$

    $[H^+] = 10^{-8.3} = 5.0 \times 10^{-9}$ M

4. In order to prepare a 0.1 M phosphate buffer with a pH of 7.2, use the Henderson-Hasselbalch equation to calculate the ratios of the salt and acid, where the salt is "A⁻" and the acid is "HA":

    $$pH = pK_a + \log \frac{[A^-]}{[HA]}$$

    From a table of ionization constants, choose the phosphate conjugate acid base pair with a pK_a closest to 7.2:

    $$H_2PO_4^- \rightleftharpoons H^+ + HPO_4^{-2}$$

    $$pK_a = 7.2$$

    Substituting these values into the equation gives:

    $$7.2 = 7.2 + \log [A^-]/[HA]$$

    $$0 = \log [A^-]/[HA]$$

    $$10^0 = 1 = [A^-]/[HA]$$

    $$[A^-] = [HA] \qquad \text{(equation 1)}$$

    The concentrations of the conjugate base and acid must be equal. We also know that the total concentration of the phosphate buffer is 0.1 M. Therefore,

    $$[A^-] + [HA] = 0.1 \text{ M} \qquad \text{(equation 2)}$$

    Using simultaneous equations (i.e., substituting the value for [HA] from equation 1 into equation 2) to determine the concentrations of the salt and acid for this particular buffer solution gives:

    $$[A^-] + [A^-] = 0.1 \text{ M}$$

    $$2 [A^-] = 0.1 \text{ M}$$

    $$[A^-] = 0.05 \text{ M}$$

    Taking this value and inserting it into (equation 1) gives

    $$[HA] = 0.05 \text{ M}$$

To prepare this buffer, place 0.05 mol each of the salt and acid in a 1 L volumetric flask and dilute with water to the 1 L mark.

6. Osmolarity = Molarity x the number of ions produced. Na_3PO_4 dissociates into four ions. Assuming 85% ionization, the osmolarity of a 1.3 M solution of Na_3PO_4 would therefore be 1.3 x 4 x 0.85 = 4.4

8. a. water and ammonia – hydrogen bonds

   b. lactate and ammonium ion – ionic interactions

   c. benzene and octane – van der Waals forces

   d. carbon tetrachloride and chloroform – van der Waals forces

   e. chloroform and diethyl ether – van der Waals forces

10. Arrows indicate atoms that would be involved in hydrogen bonding.

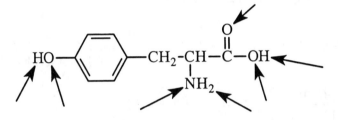

12. b, c and d all would be expected to have a dipole moment.

14. Carbon dioxide is present in the blood in sufficient quantities to make it effective as a buffer. Phosphate concentration in blood is too low to be an effective buffer. In cells the phosphate concentration is much higher, and it can therefore act as an effective buffer.

16. b, c, and e are all weak acids because they are only partially ionized. a and d are strong acids (a is hydrochloric acid and d is nitric acid).

18. Hyperventilation drives the transfer of carbon dioxide from the blood. This process, which shifts the following equilibrium to the left, consumes protons thereby making the blood more alkaline.

$$CO_2 + H_2O \rightleftharpoons H_2CO_3 \rightleftharpoons HCO_3^- + H^+$$

20. No. The carbonic acid and carbonate react to produce bicarbonate. It is possible to have either a buffer system of carbonic acid and bicarbonate or a buffer system of bicarbonate and carbonate.

22. pH=4.48

First, identify the weak acid and its conjugate base. Sodium hydrogen ascorbate would give Na+ and HAscorbate– in solution. HCl is a strong acid that would react with the HAscorbate⁻ to form H_2Ascorbate. So, the weak acid would be H_2Ascorbate and the conjugate base would be HAscorbate⁻.

The number of moles of HAscorbate⁻ initially is (300 mL)(0.25 M) = 75 mmol

The number of moles of HCl added is (150 mL)(0.2 M) = 30 mmol

It's a good assumption that HCl will react completely with the HAscorbate⁻, so that would give: 75 mmol HAscorbate⁻ – 30 mmol reacted with HCl

= 45 mmol HAscorbate⁻ left over and 30 mmol of H_2Ascorbate formed.

We also need to calculate the pK_{a1} from the K_{a1}:

$$pK_{a1} = -\log K_{a1} = -\log (5 \times 10^{-5}) = 4.30$$

Now, use the Henderson-Hasselbalch equation: $pH = pK_a + \log \dfrac{[\text{HAscorbate}^-]}{[H_2\text{Ascorbate}]}$

$$pH = 4.30 + \log \frac{[45 \text{ mmol}]}{[30 \text{ mmol}]} = 4.30 + 0.18 = \mathbf{4.48}$$

A pH of 4.48 makes sense because 4.48 is more basic than 4.30, the pK_a, and we have more base present than acid. Note that this method included a shortcut - mmol rather than molarity was used. This is valid because the total volume is the same and would cancel out. To use molarity, divide both the numerator (0.0045 moles) and the denominator (0.0030 moles) by the total volume, 0.450 L (300 mL+150 mL).

CHAPTER 3: ANSWERS TO EVEN-NUMBERED THOUGHT QUESTIONS

2. The regular crystal lattice of the ice crystal is more open than the tightly hydrogen-bound liquid water. If ice were more dense than water, ice formed in lakes and oceans would sink to the bottom. Eventually, only a narrow layer at the surface would be liquid. This environmental condition is incompatible with life (for most aquatic life, and they would not be able to survive).

4. The blood is so highly buffered by the bicarbonate buffer and the large amounts of blood proteins that under normal physiological conditions the transport of weak acids in the blood does not appreciably change its pH. For example, in the presence of bicarbonate, any acid that ionizes produces carbon dioxide (which is exhaled). The pH of the blood then remains virtually unchanged.

$$HCO_3^- + H^+ \rightleftharpoons H_2CO_3 \rightleftharpoons CO_2 + H_2O$$

6. In a liquid the molecules must be free to move over one another. In the gelatin solution, each water molecule hydrogen bonds with two segments of the protein,

locking the protein chains and the water together. Because the water molecules are no longer able to move freely, the mixture becomes semi-rigid.

8. No. The structure of cells is based on the phase separation of hydrophobic and hydrophilic substances. The function of the cell membrane is possible only because lipids are insoluble in water. If water dissolved every molecule, living organisms would not be able to create a barrier (membranes!) between themselves and their surroundings, and living organisms would not be possible.

10. The conversion of glycogen to glucose creates an increase in osmotic pressure, and water would flow into the cell. To offset this rise in osmotic pressure, ions such as sodium and potassium are pumped out of the cell. These ions would be followed by water, thus restoring cell volume. (See Special Interest Box 3.1: *Cell Volume Regulation and Metabolism*, on page 79.)

12. Water weakens ionic interactions by forming a solvation sphere around each ion. As the distance between the cation and the anion increase, the attractive force between these ions would decrease. (See Figure 3.9 on page 72.) In other words, the polar water molecules crowd around the ions, interacting with them and weakening their interactions with other ions.

ANSWERS TO *PRACTICE PROBLEMS TO PROMOTE PERFECTION*

1. Describe how to prepare 1.0 L of a 0.30 M acetate buffer with a pH of 5.4. What ratio of acetate salt to acetic acid would you use? (The pK_a of acetic acid is 4.76.)

$$pH = pK_a + \log[A^-]/[HA]$$

$$5.4 = 4.76 + \log[A^-]/[HA]$$

$$0.64 = \log[A^-]/[HA]$$

$$[A^-]/[HA] = 4.4 = \text{ratio of acetate salt to acetic acid}$$

$$[A^-] = (4.4)[HA]$$

Substitute this equation for [A⁻] into the total buffer concentration equation:

$$\text{Total buffer concentration} = \quad [A^-] + [HA] = 0.30 \text{ M}$$

$$(4.4)[HA] + [HA] = 0.30 \text{ M}$$

$$(5.4)[HA] = 0.30 \text{ M}$$

$$\mathbf{[HA] = 0.056 \text{ M}}$$

Substitute this value for [HA] into the total buffer concentration equation:

$$[A^-] + 0.056 \text{ M} = 0.30 \text{ M}$$

$$[A^-] = 0.24 \text{ M}$$

To prepare the buffer, mix 0.24 moles of acetate salt with 0.074 moles of acetic acid in a volumetric flask and dilute to the mark with water. (To be more specific, dissolve 20 grams of sodium acetate in water in a 1-L volumetric flask. Add 74 mL of 1 M acetic acid. Dilute to the 1-L mark with freshly distilled water.)

2. a. millimoles of acetic acid = (40.0 mL)(1.00M) = 40.0 mmol

millimoles of acetate = (2.00 g)/(82.0g/mol) = 24.4 mmol

pH = pK_a + log[A–]/[HA] = 4.76 + log(24.4/40.0) = **4.55**

(Note that we don't need to use the total volumes to get the correct answer. If you calculated molarity, the equation would be:

pH = 4.76 + log([0.0813M A–]/[0.133M HA]) = 4.55

b. Total buffer concentration = (40.0 mmoles + 24.4 mmoles)/300 mL = 0.215 M

c. HCl: (10 mL)(1 M) = 10 mmol HCl, which would react with the acetate to form acetic acid

acetate: 24.4 mmol – 10 mmol (reacted with HCl) = 14.4 mmol acetate

acetic acid: 40.0 mmol + 10 mmol (formed) = 50 mmol acetic acid

pH = 4.76 + log(14.4/50) = 4.76 – .54 = **4.22**

It's still a good buffer! And, if 10 mmol HCl were in a total volume of 310 mL, the pH would be **1.5**. Buffers work!

3. Distilled water that has been exposed to air for any length of time will probably contain dissolved CO_2, which would lower the pH.

$$CO_2 + H_2O \rightleftharpoons H_2CO_3 \rightleftharpoons H^+ + HCO_3^-$$

4. First, calculate molarity of the fructose: 27 grams/180 g/mol = 0.15 moles

0.15 moles/0.240 L = 0.625 M

T = 37 + 273 = 310 K

π = iMRT = (1)(0.625 M)(0.082)(310 K) = 16 atm

The molarity of sucrose would be: 27 grams/342 g/mol = 0.0790 moles

0.0790 moles/0.240 L = 0.329 M

π = iMRT = (1)(0.329 M)(0.082)(310 K) = 8.4 atm

5. osmolarity = iM = (2)(4.53 x 10^{-3} M) = 9.06 x 10^{-3} M

osmotic pressure = π = iMRT = (2)(4.53 x 10^{-3} M)(0.082)(310 K) = 0.23 atm

6. π = iMRT, R = 0.082 L·atm/K·mol

T = 25° C + 273 = 298 K

i = 1 (for a nonelectrolyte)

M = moles/liter = moles/(0.100 L)

First, use π = iMRT to solve for number of moles:

0.465 atm = (1)(moles/0.100L)(0.082 L·atm/K·mol)(298 K)

1.90 x 10^{-3} moles

molecular weight = grams/moles = (0.200 g)/(1.90 x 10^{-3} moles) = 105 g/mol

Energy

INTRODUCTION: THERMODYNAMICS AND BIOENERGETICS

enthalpy (*H*), entropy (*S*), free energy (*G*)

THERMODYNAMICS

Open vs. closed systems

State functions - values are independent of path

Enthalpy, entropy, and free energy are state functions, but work and heat are not.

Energy is the capacity to do work.

Energy is exchanged (between system and surroundings) as work and/or heat.

First Law of Thermodynamics

Energy can neither be created nor destroyed, but it can be transformed from one form into another.

Enthalpy ($\Delta H_{\text{reaction}} = H_{\text{products}} - H_{\text{reactants}}$); exothermic vs. endothermic processes

Since enthalpy is a state function, the reaction mechanism (i.e., *how* you get from the reactants to the products) doesn't affect ΔH. That means that we can calculate the standard enthalpy of any reaction by summing up the ΔH_f° (the standard enthalpy of formation per mole) of the products and subtracting the sum of the ΔH_f° of the reactants:

$$\Delta H^\circ = \Sigma \Delta H_{f\,(\text{products})}^\circ - \Sigma \Delta H_{f\,(\text{reactants})}^\circ$$

Don't forget to take the stoichiometry of the reaction into consideration. You'll need to multiply the standard enthalpy of formation of each molecule by its coefficient in the chemical reaction.

Second Law of Thermodynamics

The disorder (entropy, ΔS) of the universe always increases.

Spontaneous reactions or processes are exergonic – energy is released.

Systems *can* spontaneously become more ordered (decrease entropy) IF the surroundings become more disordered (increase entropy) and the overall disorder (entropy) of the universe *increases*.

FREE ENERGY (ΔG = GIBBS FREE ENERGY CHANGE)

$\Delta G = \Delta H - T\Delta S$ *(at constant temperature and pressure)*

Exergonic (spontaneous) vs. endergonic (nonspontaneous) processes

The sign of ΔG allows us to predict whether or not a chemical reaction can occur.

If: then the reaction is:

$\Delta G = -$ spontaneous (exergonic, favorable)

$\Delta G = +$ nonspontaneous (endergonic, not favorable)

$\Delta G = 0$ at equilibrium (no change)

Consider a general reaction, A+B → C+D. If the ΔG for this reaction is negative (energy is released and it's spontaneous), then the reverse reaction, C+D → A+B, has to be nonspontaneous (ΔG is positive). Since ΔG is a state function, the magnitude of ΔG in forward and reverse reactions is the same. So, when the direction of a chemical reaction is reversed, the sign of ΔG is also reversed.

When $\Delta G =$ zero, there is no net change in the chemical reaction. The rate forward (A+B → C+D) equals the rate of the reverse reaction (C+D → A+B), and the reaction is at equilibrium.

Standard Free Energy Changes: $\Delta G°$

Since ΔG depends on temperature, pressures, and concentrations, there needs to be some reference point. The *standard free energy*, $\Delta G°$, is defined as ΔG at standard state conditions: 25°C, 1 atm pressure, and 1 M reactant concentration. However, nearly all biochemical reactions occur in dilute, aqueous mixtures, so biochemists have developed their own reference point, $\Delta G°'$, which is $\Delta G°$ at pH = 7.[1]

ΔG = free energy change of a reaction under actual conditions

$\Delta G°$ = free energy change at standard state conditions: 25°C, 1.0 atm, 1.0 M

$\Delta G°'$ = free energy change at biochemical standard state conditions: $\Delta G°$ at pH 7.

So, the symbol (°) indicates standard state conditions, and a prime (′) indicates biochemical standard state conditions.

ΔG is related to the reference $\Delta G°$ (or $\Delta G°'$) by the following equation:[2]

$$\Delta G = \Delta G° + RT \ln \frac{[C]^c[D]^d}{[A]^a[B]^b} \quad \text{for the reaction: } a\,A + b\,B \rightarrow c\,C + d\,D$$

(R = 8.315 J/mol K and T = °C+273)

At equilibrium, $\Delta G = 0$, so: $\Delta G° = -RT \ln K_{eq}$

K_{eq} is the equilibrium constant: $K_{eq} = \dfrac{[\text{products}]_{\text{at equilibrium}}}{[\text{reactants}]_{\text{at equilibrium}}} = \dfrac{[C]^c[D]^d}{[A]^a[B]^b}$

The sign and value of $\Delta G°$ indicates the direction and magnitude of a particular reaction at equilibrium (and under standard-state conditions). Also, the K_{eq} can be calculated for a given $\Delta G°$.

---

[1] Note that if [H^+] does not affect the chemical reaction, then $\Delta G° = \Delta G°'$.

[2] Equilibrium constants are really defined for "activities" rather than "concentrations." However, concentrations are simpler to use and adequate for our purposes.

EXAMPLE:

Calculate K_{eq} for the hydrolysis of ATP, given the following:

$$ATP + H_2O \rightarrow ADP + P_i \qquad \Delta G° = -30.5 \text{ kJ/mol and } K_{eq} = \frac{[ADP][P_i]}{[ATP]}$$

(Since water is the solvent, $[H_2O]$ is omitted from K_{eq}.)

Solution:

We can solve for K_{eq} using: $\Delta G° = -RT \ln K_{eq}$

$$-30.5 \text{ kJ/mol} = -(8.315 \text{ J/mol K})(298 \text{ K}) \ln K_{eq}$$

$$12.31 = \ln K_{eq}$$

$$K_{eq} = 2.22 \times 10^5 \text{ or } \frac{222,000}{1}$$

In this example, you can see that a negative $\Delta G°$ produces a large K_{eq}. The large K_{eq} indicates that at equilibrium (at 25°C), there are 222,000 times more product than reactant, and the reaction will proceed in the forward direction. Another way to look at this is that at equilibrium, for each molecule of ATP present, there will be 471 ADP molecules and 471 P_i molecules (since $[ADP][P_i]$ =222,000 and the square root of 222,000 is 471).

What if the $\Delta G°$ was + 30.5 kJ/mol? The K_{eq} would be 4.5 x 10^{-6} or $\frac{1}{222,000}$.

$$ATP + H_2O \rightarrow ADP + P_i \qquad \Delta G° = -30.5 \text{ kJ/mol, spontaneous}$$

$$ADP + P_i \rightarrow ATP + H_2O \qquad \Delta G° = +30.5 \text{ kJ/mol, nonspontaneous}$$

Although a reaction with a positive $\Delta G°$ is nonspontaneous, that same reaction may have a negative ΔG under different (nonstandard-state) conditions, e.g., in a living cell. The actual direction of a reaction can be altered by changing the concentration of products or reactants. This is described by the general equation:

$$\Delta G = \Delta G° + RT \ln \frac{[products]}{[reactants]} \qquad or \qquad \Delta G = \Delta G° + RT \ln \frac{[C]^c[D]^d}{[A]^a[B]^b}$$

EXAMPLE:

$$\text{glucose-1-phosphate} \rightarrow \text{glucose-6-phosphate} \qquad \Delta G° = -7.1 \text{ kJ/mol}$$

From the negative $\Delta G°$, we already know that the reaction proceeds forward to form glucose-6-phosphate under biochemical standard state conditions. But, can this reaction be reversed? Intuitively, if we were to add a high concentration of glucose-6-phosphate, we would think that this reaction could be "pushed" backwards. What if the concentration of glucose-6-phosphate was 100 mM and the concentration of glucose-1-phosphate was 1 mM? In which direction would the reaction proceed?

Let's use the equation: $\quad \Delta G = \Delta G° + RT\, ln\, \dfrac{[glucose - 6 - phosphate]}{[glucose - 1 - phosphate]}$

[glucose-6-phosphate] = 100 mM

[glucose-1-phosphate] = 1 mM

$\Delta G = -7100 \text{ J/mol} + (8.315 \text{ J/mol K}) (298 \text{ K})\, ln\, \dfrac{100 \text{ mM}}{1 \text{ mM}}$

$\Delta G = + 4.3 \text{ kJ/mol}$

The positive ΔG tells us that this reaction will proceed in the opposite direction as written. So, by increasing the concentration of the product, we would be able to reverse the direction of the reaction.

As you can see, the direction of a reaction under actual cellular conditions depends not only on the $\Delta G°$, but also on the concentrations of reactants and products.

Coupled Reactions

Coupled reactions allow the cell to harness the energy produced by catabolism.

EXAMPLE:

Compare the reactions for the hydrolysis of PEP (phosphoenolpyruvate) and for the formation of ATP:

$$PEP + H_2O \quad \rightarrow \quad pyruvate + P_i \qquad\qquad \Delta G°' = -61.9 \text{ kJ/mol}$$

$$ADP + P_i \quad \rightarrow \quad ATP + H_2O \qquad\qquad \Delta G°' = +30.5 \text{ kJ/mol}$$

Note that the hydrolysis of PEP is spontaneous ($\Delta G°'$ is negative, and the reaction proceeds in the forward direction as written), but the reaction to form ATP is not ($\Delta G°'$ is positive). To capture some of the energy of this PEP hydrolysis, consider coupling (adding) these two reactions:

$$PEP + H_2O \quad \rightarrow \quad pyruvate + P_i \qquad\qquad \Delta G°' = -61.9 \text{ kJ/mol}$$
$$ADP + P_i \quad \rightarrow \quad ATP + H_2O \qquad\qquad \Delta G°' = +30.5 \text{ kJ/mol}$$
_____
$$PEP + ADP \quad \rightarrow \quad pyruvate + ATP \qquad\qquad \Delta G°' = -31.4 \text{ kJ/mol}$$

When we add these reactions, we also add the individual $\Delta G°'$ values to obtain the $\Delta G°'$ of the overall reaction: $\quad \Delta G°'_{overall} = \Delta G°'_{reaction\ 1} + \Delta G°'_{reaction\ 2}$ (This is valid because ΔG is a state function.)

The overall reaction is still thermodynamically favorable ($-\Delta G°'$), but we used some of the free energy to "drive" an unfavorable chemical reaction. Some of the energy released from the hydrolysis of PEP (phosphoenolpyruvate) was captured by ADP to form ATP.

This captured energy can be released later (when it's needed) by hydrolyzing ATP to regenerate ADP. For example, consider the phosphorylation of glucose:

$$glucose + P_i \quad \rightarrow \quad glucose\text{-}6\text{-}phosphate + H_2O \qquad\qquad \Delta G°' = +13.8 \text{ kJ/mol}$$

When this reaction is coupled to ATP hydrolysis, the overall $\Delta G^{\circ\prime}$ is negative. The energy from ATP hydrolysis drives this endergonic reaction forward.

$$\text{glucose} + P_i \rightarrow \text{glucose-6-phosphate} + H_2O \qquad \Delta G^{\circ\prime} = +13.8 \text{ kJ/mol}$$

$$\underline{ATP + H_2O \rightarrow ADP + P_i \qquad\qquad\qquad\qquad \Delta G^{\circ\prime} = -30.5 \text{ kJ/mol}}$$

$$\text{glucose} + ATP \rightarrow \text{glucose-6-phosphate} + ADP \qquad \Delta G^{\circ\prime} = -16.7 \text{ kJ/mol}$$

Since cells use ATP synthesis and hydrolysis to store and use chemical energy, ATP is often referred to as the "energy currency" of the cell.

In summary, a reaction that is not spontaneous can be driven forward *if* it's coupled with a spontaneous reaction *and* the overall ΔG is negative. In other words, the nonspontaneous reaction needs the energy from another reaction to drive it forward.

How can we get a nonspontaneous reaction to go forward (assuming constant temperature)?

- Couple it with a spontaneous reaction that will supply enough energy to give an overall negative ΔG, or

- Change the relative concentrations of reactant and product so that the actual ΔG will be negative.

The Hydrophobic Effect Revisited

The hydrophobic effect explains why nonpolar molecules aggregate in water, why micelles and bilayers form, and why proteins fold. Although these processes seem to result in a decrease in entropy (i.e., an increase in order), the overall entropy (including that of the surrounding water molecules) is higher.[3]

THE ROLE OF ATP

ATP is produced using the energy released by breaking down food molecules (catabolism) and by the light reactions of photosynthesis. Hydrolysis of ATP releases 30.5 kJ/mole of energy that is used to drive endergonic processes such as:

1. biosynthesis (anabolic pathways)
2. active transport of substances across cell membranes
3. mechanical work (e.g., muscle contraction)

Why is ATP hydrolysis so exergonic? The products are more stable than the reactants, because the final products:

1. have less electrostatic repulsion (of negative charges)
2. have more resonance structures than the reactants[4]
3. are more easily solvated than the reactants

---

[3] This is an oversimplification. Surface tension effects, van der Waals interactions, and the ordered arrangement of the water molecules immediately adjacent to a nonpolar molecule should also be taken into consideration. In general, at biological temperatures, these processes are driven by an increase in entropy (disorder) of the surroundings.

[4] Of course, this assumes that the H^+ of ADP dissociates.

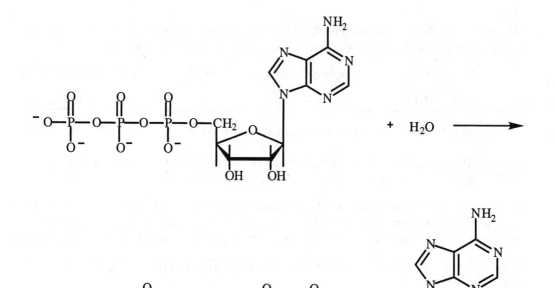

Phosphate group transfer potential is the tendency of a phosphate-containing molecule to hydrolyze, resulting in its phosphate group being released as HPO_4^{2-} or transferred to another molecule. The greater the phosphate group transfer potential, the more energy is released when a phosphate group is hydrolyzed (and the more stable a molecule would be *without* its phosphate group).

Since ATP has an intermediate phosphate group transfer potential, it can take a phosphate group from a higher-energy compound and transfer it to a lower-energy compound. For instance, recall the two examples given above for coupled reactions:

$$PEP + ADP \quad \rightarrow \quad pyruvate + ATP \qquad\qquad \Delta G^{\circ\prime} = -31.4 \text{ kJ/mol}$$

$$glucose + ATP \quad \rightarrow \quad glucose\text{-}6\text{-}phosphate + ADP \qquad \Delta G^{\circ\prime} = -16.7 \text{ kJ/mol}$$

Essentially, a phosphate group could be transferred from PEP to ATP, then from ATP to glucose, because both of these coupled reactions are spontaneous. PEP has a higher phosphate group transfer potential than ATP, and ATP has a higher phosphate group transfer potential than glucose-6-phosphate.

REDOX IN THE DEEP: PHOTOAUTOTROPHS, LITHOTROPHS, CHEMOLITHOTROPHS

THERMODYNAMICS VS. KINETICS

Thermodynamics tell us whether or not a reaction will be spontaneous. If a reaction is "spontaneous," there will be more products than reactants when the reaction is in equilibrium. That is, the reaction will proceed forward until it reaches equilibrium. Spontaneous reactions are exergonic – they release energy.

Kinetics tell us how *fast* the reaction will go.

Consider the combustion of glucose: $C_6H_{12}O_6 + 6\,O_2 \rightarrow 6\,CO_2 + 6\,H_2O$.

Glucose reacts with oxygen to form carbon dioxide and water. Thermodynamics tell us that this reaction is spontaneous and highly exothermic, but kinetics tell us that the rate at room temperature is incredibly slow. That's why glucose won't simply burst into flames when air hits it.

So, the chemical term "spontaneous" has a different meaning from the common dictionary definition. Glucose will *spontaneously* combust, but it will not *suddenly* combust (unless enough energy is applied to overcome its activation energy, and/or the activation energy is lowered with the help of a catalyst or enzyme).[5]

THERMODYNAMICS	KINETICS
Spontaneity: *Can* a reaction happen?	Rate: *How fast* will a reaction happen?
Gibbs free energy change, ΔG	E_a, activation energy
exergonic vs. endergonic	k (rate constant)
enthalpy, ΔH, and entropy, ΔS	order of a reaction
$\Delta G = \Delta H - T\Delta S$	depends on mechanism (path)
equilibrium	Enzymes - biochemical catalysts that reduce the activation energy and cause a reaction to go faster. Enzymes do not affect the thermodynamics of a reaction.
K_{eq} (equilibrium constant)	
Chapter 4: Energy	Chapter 6: Enzymes

AFTER STUDYING THIS CHAPTER, YOU SHOULD BE ABLE TO SOLVE THESE TYPES OF PROBLEMS:

- Calculate $\Delta H_{reaction}$ given ΔH_f° data

- determine whether a reaction is: endothermic or exothermic, exergonic or endergonic, spontaneous or nonspontaneous

- use $\Delta G = \Delta G^{\circ} + RT\,ln\dfrac{[products]}{[reactants]}$ to determine whether or not a reaction will proceed at particular concentrations (i.e., is it spontaneous under different cellular conditions?)

- Coupled reactions: Determine whether the overall reaction will be spontaneous using the equation: $\Delta G^{\circ\prime}_{overall} = \Delta G^{\circ\prime}_{reaction\ 1} + \Delta G^{\circ\prime}_{reaction\ 2}$

- Coupled reactions: Choose a reaction that will provide enough energy to drive a non-spontaneous reaction.

---

[2] "Spontaneous combustion" as defined by popular usage is beyond the scope of this discussion.

EQUATIONS

$$\Delta H^\circ = \Sigma \Delta H_f^\circ{}_{(products)} - \Sigma \Delta H_f^\circ{}_{(reactants)}$$

$$\Delta G = \Delta H - T\Delta S$$

$$K_{eq} = \frac{[products]_{at\ equilibrium}}{[reactants]_{at\ equilibrium}} = \frac{[C]^c[D]^d}{[A]^a[B]^b} \quad \text{(for the reaction } aA + bB \Leftrightarrow cC + dD)$$

(For the Keq term, remember to use exponents if the coefficients in the chemical equation are greater than one.)

$$\Delta G = \Delta G^\circ + RT\ ln\ \frac{[products]}{[reactants]} \quad R = 8.315 \text{ J/mol K and } T = {}^\circ C + 273$$

(Be careful with J vs. kJ, since R contains joules and ΔG is typically in kilojoules. 1000 J = 1 kJ)

$$\Delta G^\circ = -RT\ ln\ K_{eq}$$

For coupled reactions: $\Delta G^{\circ\prime}{}_{overall} = \Delta G^{\circ\prime}{}_{reaction\ 1} + \Delta G^{\circ\prime}{}_{reaction\ 2}$

FOOD FOR THOUGHT:

What's wrong with the following statement?
    ATP releases energy when its high-energy phosphate bond is broken.

Answer:

1. The bond is not simply "broken." It's hydrolyzed (cleaved with the addition of water). Some bonds are broken, and others are formed.

2. The term "high-energy bond" suggests that the bond is unstable. Stability really refers to its ability to participate in reactions as opposed to the magnitude of the bond energy. The phosphoanhydride bond of ATP is actually more stable relative to compounds that have a higher phosphate group transfer potential.

CHAPTER 4: ANSWERS TO EVEN-NUMBERED REVIEW QUESTIONS

2. State functions are those that are independent of path. Entropy, enthalpy, and free energy are all path independent.

4. Given that the ionization constant for formic acid is 1.83×10^{-4}, the $\Delta G°$ for the reaction would be calculated as follows:

$$\Delta G° = - RT \ln K_{eq}$$

$$\Delta G° = - (8.315 \text{ J/mol K}) (298 \text{ K}) \ln (1.8 \times 10^{-4})$$

$$\Delta G° = - (8.315) (298) (-8.62)$$

$$\Delta G° = + 21366 \text{ J/mol or } 21.3 \text{ kJ/mol}$$

6. d: Reaction rates cannot be determined from energy values.

8. Under standard conditions the following statements are true: b, d, and e.

10. AMP hydrolysis involves cleavage of an ester bond and therefore releases the least energy. Hydrolysis of the other phosphate linkages involves either the hydrolysis of an anhydride or enol bond.

CHAPTER 4: ANSWERS TO EVEN-NUMBERED THOUGHT QUESTIONS

2. $\Delta G° = -RT \ln K_{eq}$

$-16{,}700 \text{ J/mol} = -(8.315 \text{J/mol K})(298 \text{ K}) \ln K_{eq}$

$-16{,}700 = - 2478 \ln K_{eq}$

$16{,}700/2478 = \ln K_{eq}$

$6.74 = \ln K_{eq}$

$K_{eq} = 851$

4. In the case of endothermic solutions, the enthalpy may be negative but the entropy is sufficiently positive to make the overall $\Delta G°$ favorable.

6. ATP has an intermediate phosphate group transfer potential. This makes it possible for ATP to serve as a carrier of phosphoryl groups from high energy compounds to those of lower energy.

8. $\Delta H = -88 \text{ kJ/mol}$ $\qquad \Delta S = 0.3 \text{ kJ/mol K} \qquad T = 298 \text{ K}$

$\Delta G = \Delta H - T\Delta S$

$\Delta G = (-88 \text{ kJ/mol}) - (298)(0.3 \text{ kJ/mol K})$

$= - 88 - 89.4 = -177.4 \text{ kJ/mol}$

$\Delta G° = - RT \, ln \, K_{eq}$

$(-177.4 \text{ kJ/mol})(1000 \text{ J/kJ}) = - (8.315 \text{ J/mol K}) (298 \text{ K}) \, ln \, K_{eq}$

$71.4 = ln \, K_{eq}$ and $K_{eq} = 1.22 \times 10^{31}$

The K_{eq} value indicates that the denaturation of the protein is virtually irreversible.

10. When magnesium ions coordinate with the phosphate groups of ATP, the repulsion between adjacent oxygen anions is decreased. Consequently, ATP is stabilized and the free energy of hydrolysis is reduced. (Conditions that stabilize the reactant (ATP) would make the ΔG less negative.)

Amino Acids, Peptides, and Proteins

INTRODUCTION

amino acids, amino acid residues, peptides, polypeptides, and proteins

AMINO ACIDS

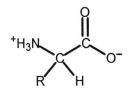

- 20 standard amino acid structures and abbreviations

- standard vs. nonstandard amino acids

- amphoteric; at pH=7, amino acids exist as zwitterions

Amino Acid Classes:

- Nonpolar, neutral (aromatic, aliphatic, and sulfur-containing side chains)

- Polar, neutral (side chains contain –OH or an amide)

- Acidic (side chains contain –COOH)

- Basic (side chains contain an N that can gain an H^+)

Biologically Active Amino Acids

- Chemical messengers: neurotransmitters (GABA, serotonin, melatonin), hormones (thyroxine, indole, acetic acid)

- Precursors to complex N-containing molecules

- Metabolic intermediates

Modified Amino Acids in Proteins

- Amino acid derivatives

- Examples: carboxylation, hydroxylation, phosphorylation

Amino Acid Stereochemistry - affects structure and function of proteins

- chiral carbons, stereoisomers, enantiomers, optical isomers

- dextrorotary (+) vs. levorotary (–)

- L-amino acids; L- designation refers to similarity with L-glyceraldehyde at the chiral carbon furthest from the C=O

Titration of Amino Acids

- Draw the amino acid in its most acidic form.
- Draw the amino acid structure that occurs when it reacts with one OH⁻. Then, draw structures that react with further OH⁻ until the amino acid is in its most basic form. (Groups with the lowest pK_a values donate their H⁺ ions first.)
- Isoelectric point (pI) of amino acids and peptides

Amino Acid Reactions

- Peptide bond formation
    - peptide bond has partial double-bond character due to resonance, so the peptide bond is rigid and planar
    - draw peptides from N-terminal to C-terminal
    - name peptides (amino acid sequence from N-terminal to C-terminal)
    - α-carbon (C_α) is next to the peptide-bonded C=O
    - ψ = rotation around the C_α–N bond; ϕ = rotation around the C_α–C bond
- Cysteine oxidation to form disulfide bridges
    - covalent S–S bond formed by the oxidation of two cysteine R groups:

        $$R–SH + HS–R \Leftrightarrow R–S–S–R \text{ (cystine)}$$

    - can occur within a chain or between two separate chains

PEPTIDES

Examples of peptide functions with specific peptides that perform these functions:

Reducing agent

glutathione (GSH, where "SH" indicates a cysteine R group)

$$2\,GSH + H_2O_2 \rightarrow GSSG + 2\,H_2O \text{ (Note the disulfide bond formation.)}$$

Appetite control:

α-melanocyte stimulating hormone, cholecystokinin, galanin, neuropeptide Y

Blood pressure control: vasopressin, atrial natriuretic factor

Pain perception:

opioid peptides (Met-enkephalin, Leu-enkephalin) vs. Substance P, bradykinin

PROTEINS: CLASSIFIED BY FUNCTION, SHAPE, OR COMPOSITION

PROTEIN FUNCTIONS: catalysis, structure, movement, defense, regulation, transport, storage, stress response

PROTEIN SHAPES: fibrous proteins vs. globular proteins

PROTEIN COMPOSITIONS: simple vs. conjugated

Conjugated protein = simple protein + prosthetic group. *Holoproteins Have* their prosthetic groups. In *Apoproteins*, prosthetic groups are *Absent*.

Examples of conjugated proteins: glycoproteins, lipoproteins, metalloproteins, phosphoproteins, hemoproteins

Protein Structure

PRIMARY STRUCTURE = AMINO ACID SEQUENCE

Homologous polypeptides

Invariant vs. variable residues; mutations that change the amino acid sequence and their connection to evolution and molecular diseases

Homozygous vs. heterozygous

SECONDARY STRUCTURE = REPEATING PATTERNS OF LOCALIZED STRUCTURE

α-helix: right-handed helix, 3.6 residues per turn, pitch = 54 nm

Hydrogen bonds between N–H and C=O are four residues apart, and R groups extend *outward* from the helix.

Amino acids that are incompatible with the α-helix: Gly, Pro, and sequences with large numbers of charged and/or bulky R groups.

β-pleated sheet; β-strand; parallel vs. antiparallel

Stabilized by H-bonds between N-H and C=O of adjacent chains

Each β-strand is fully extended. Antiparallel is more stable than parallel because the hydrogen bonds between chains are more direct (*colinear* and shorter).

Supersecondary structures: $\beta\alpha\beta$ unit, β-meander, $\alpha\alpha$-units, β-barrel, Greek key

TERTIARY STRUCTURE AND PROTEIN FOLDING OF GLOBULAR PROTEINS

Features: 1. amino acids that are far apart in the primary structure may be close together once folded; 2. globular proteins are compact; 3. large globular proteins often contain domains - compact units with specific functions (examples: EF hand binds Ca^{2+}; leucine zipper and zinc finger domains found in DNA-binding proteins)

Stabilized by hydrophobic interactions, electrostatic interactions (salt bridges), hydrogen bonds, covalent bonds (disulfide bridges)

QUATERNARY STRUCTURE – SEVERAL SUBUNITS (POLYPEPTIDE CHAINS)

Oligomers and protomers; why multisubunit proteins are common

Noncovalent and covalent interactions hold the subunits in place. The most important interaction is the hydrophobic effect.

Covalent crosslinks: disulfide bridges, desmosine and lysinonorleucine

Allostery, ligand binding, allosteric transitions, effectors or modulators

PROTEIN DYNAMICS AND FLEXIBILITY in a protein's 3-D structure; why flexibility in a protein's 3-D structure is essential to most protein functions

PROTEIN DENATURATION – loss of 3-D structure (but peptide bonds are NOT broken)

Causes include strong acids or bases, organic solvents, detergents, reducing agents, salt concentration (salting out), heavy metal ions, temperature changes, mechanical stress

Fibrous Proteins Typically Serve Structural Functions

High proportions of regular secondary structures; rodlike or sheetlike shapes

Examples: α-keratin, collagen, silk fibroin

Globular Proteins Typically Serve Dynamic Functions

- Function: usually involves binding ligands or large biomolecules that induce a conformational change "linked to a biochemical event."

- Myoglobin (in skeletal and cardiac muscle), hemoglobin (in red blood cells)

- Heme protein decreases heme's affinity for O_2 and protects Fe^{2+} from irreversibly oxidizing to Fe^{3+} (hematin), so O_2 can bind reversibly (like a post-it note that can be placed or removed when needed).

- Fetal hemoglobin (HbF) has greater affinity for O_2 than HbA, maternal hemoglobin. (How else could a growing baby receive O_2 from the mother?)

- Myoglobin has a greater affinity for oxygen than hemoglobin. So, oxygen moves from blood to muscle, and myoglobin only gives up its oxygen when the muscle cell's O_2 concentration is very low.

- Taut state vs. Relaxed state (T, R states)

- Cooperative binding: The binding of the first ligand (example - O_2 to hemoglobin) causes a conformational change that facilitates binding of more ligands (3 more O_2 molecules to hemoglobin).

- Bohr effect: Dissociation of O_2 from hemoglobin is enhanced at lower pH. Why? Higher levels of CO_2 cause higher $[H^+]$, since $CO_2 + H_2O \rightarrow HCO_3^-$ and H^+. H^+ stabilizes the deoxy form of hemoglobin, so it's formed faster.

HINTS FOR LEARNING THE 20 STANDARD AMINO ACIDS

- Learn them, don't just memorize them. Flash cards help. Write the name on one side and the structure on the other.

- Write out the structures for yourself. As you learn the structures, make a shorter list of the ones that continue to be troublesome for you.

- Group similar amino acids together.

- Group confusing amino acids together. For example, leucine and isoleucine are troublemakers, since leucine has the isobutyl group, not isoleucine. We're all stuck with the chore of keeping those two straight. Also, there's aspartate and glutamate, but it's helpful that aspartate has the shorter chain and also comes before glutamate in the dictionary.

- Try drawing the structures differently. For example, comparing tryptophan and histidine drawn as shown below may help you to learn these two problem children.

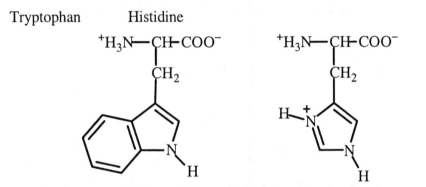

- Why is this important? References to amino acids occur throughout this course. It will be much easier for you to learn them now, so that when you learn about an enzyme that is regulated by a reaction at its tyrosine residue, you'll have a clear picture of the structure without having to look it up.

THREE-LETTER ABBREVIATIONS = FIRST THREE LETTERS *EXCEPT...*

The three-letter abbreviation for most of the amino acids is its first three letters, with the following exceptions:

Ile	=	Isoleucine	Asn =	Asparagine
Trp	=	Tryptophan	Gln =	Glutamine

HOLOPROTEIN VS. APOPROTEIN

*H*oloproteins *H*ave the prosthetic group. In *A*poproteins, the prosthetic group is *A*bsent.

AMINO ACIDS WITH IONIZABLE R GROUPS

Amino Acid	Structure at pH=7 with R group pK_a	Amino Acid	Structure at pH=7 with R group pK_a
Aspartate Asp	$^+H_3N-CH-C-O^-$, O, CH₂, C=O, O⁻ **3.86**	**Glutamate** Glu	$^+H_3N-CH-C-O^-$, O, CH₂, CH₂, C=O, O⁻ **4.25**
Tyrosine Tyr	$^+H_3N-CH-C-O^-$, O, CH₂, (phenol ring), OH **10.07**	**Lysine** Lys	$^+H_3N-CH-C-O^-$, O, CH₂, CH₂, CH₂, CH₂, NH₃⁺ **10.79**
Histidine His	$^+H_3N-CH-C-O^-$, O, CH₂, (imidazole ring N, NH) **6.0**	**Cysteine** Cys	$^+H_3N-CH-C-O^-$, O, CH₂, SH **8.33**
Arginine Arg	$^+H_3N-CH-C-O^-$, O, CH₂, CH₂, CH₂, NH, C=NH₂⁺, NH₂ **12.48**	**Unusual Functional Groups** • Guanidino: in arginine's R group • Imidazole: in histidine's R group • Phenol: in tyrosine's R group • Sulfhydryl: in cysteine's R group (It's also called a thiol.)	

THE ORGANIC CONNECTION: WHY *DON'T* SOME AMINO ACIDS ACCEPT OR DONATE AN H⁺? AND WHAT'S THE DEAL WITH HISTIDINE?

Tryptophan's nitrogen does NOT accept an H⁺, because the lone pair of electrons on the nitrogen are too busy being aromatic. (That is, if it *did* accept an H⁺, the ring would no longer be aromatic, and that's just too darn unstable to happen at normal pH values.) For the same reason, histidine's singly-bonded nitrogen (in the imidazole ring) doesn't accept an H⁺. Histidine's doubly-bonded nitrogen has a lone pair of electrons that do not contribute to making that ring aromatic, so it can accept an H⁺.

Serine's and threonine's –OH groups don't ionize within the pH range of 1-14. Tyrosine's –OH group does because it's a phenol. The resulting –O⁻ is stabilized by resonance structures around the aromatic ring.

Asparagine's and glutamine's –NH$_2$ doesn't accept an H⁺ because the lone pair on the amide nitrogen is busy being in resonance with the carbonyl beside it, just like in a peptide bond.

ISOELECTRIC POINT = pH AT WHICH THE AMINO ACID HAS NO NET CHARGE

To determine the isoelectric point:

1. Identify the acid(s) and base(s) on the amino acid or peptide. Which functional groups can happily lose or gain an H⁺? (*'Happily'* means within the pH range of 1–14.)

2. Rank the functional groups in the order in which they will lose an H⁺. Remember that the lower the pK_a, the stronger the acid, so the acid with the lowest pK_a will lose its H⁺ first.

3. Average the pK_as on either side of the isoelectric (neutral) molecule. Note: Be aware that if an acidic or basic side group is present, you *never* average *all* of the pK_as. Only average these two pK_as: the pK_a to go from a net charge of +1 to 0, and the pK_a to go from a net charge of 0 to –1.

   If it's not clear which two pK_as to average, first draw the amino acid or peptide at the lowest (acidic) pH and label each ionizable group with its pK_a. Then remove one H⁺ at a time (in order of pK_a) and draw each structure. It helps to label the reaction arrows with their pK_a values and each structure with its net charge.

The examples of alanine and glutamic acid are in your text (pp. 117-119), along with a tetrapeptide that includes lysine and aspartic acid. Also, see Review Question #4 for practice with histidine.

More pI practice problems are at the end of this section.

Remember that the lower the pK_a, the stronger the acid. Also remember that when pH=pK_a, [A⁻]=[HA]. Below the pK_a, there will be more HA. Above the pK_a, there will be more A⁻.

CALCULATING PI: LET'S LOOK AT THE EXAMPLE OF LYSINE IN MORE DETAIL.

1. Identify acidic and basic functional groups.

–COOH is a weak acid and can donate its H^+ to form –COO⁻. ($pK_a = 2.18$)

–NH₂ is a weak base and can accept an H^+ to form –NH₃⁺. ($pK_a = 8.95$)

The R group also has an –NH₂ that can accept an H^+ to form –NH₃⁺. ($pK_a = 10.07$)

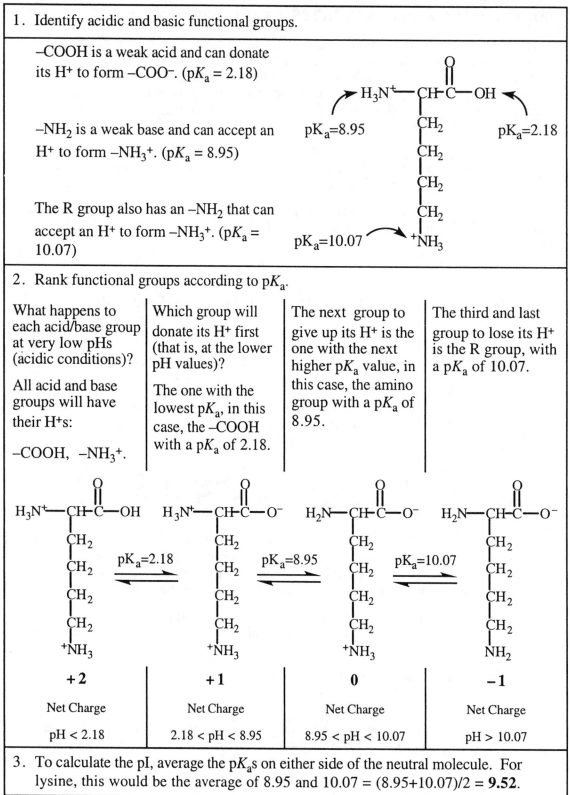

2. Rank functional groups according to pK_a.

What happens to each acid/base group at very low pHs (acidic conditions)?	Which group will donate its H^+ first (that is, at the lower pH values)?	The next group to give up its H^+ is the one with the next higher pK_a value, in this case, the amino group with a pK_a of 8.95.	The third and last group to lose its H^+ is the R group, with a pK_a of 10.07.
All acid and base groups will have their H^+s: –COOH, –NH₃⁺.	The one with the lowest pK_a, in this case, the –COOH with a pK_a of 2.18.		
+2	**+1**	**0**	**−1**
Net Charge	Net Charge	Net Charge	Net Charge
pH < 2.18	2.18 < pH < 8.95	8.95 < pH < 10.07	pH > 10.07

3. To calculate the pI, average the pK_as on either side of the neutral molecule. For lysine, this would be the average of 8.95 and 10.07 = (8.95+10.07)/2 = **9.52**.

PEPTIDE BOND

Amino acids are linked by a covalent bond between the α-amino (-NH₂) of one amino acid and the α-carboxyl (-COOH) of another amino acid. This covalent bond has a special name, called **peptide** or **amide bond**.

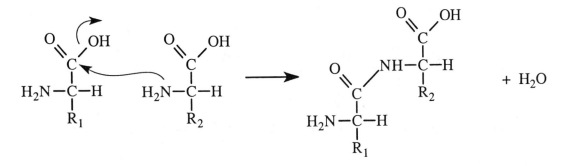

The peptide C–N bond is shorter and much more rigid than a typical carbon-nitrogen single bond. This can be explained by drawing the resonance structure for an amide bond:

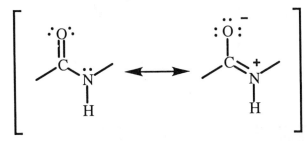

Although this resonance structure may seem unlikely because of the charge separation, it is indeed significant. Resonance explains the observed similarities of the peptide bond to a carbon-carbon double bond, namely, its rigidity and its shorter-than-expected bond length.

A carbon-nitrogen double bond is planar, or flat, and this makes the peptide bond rigid. The planar, rigid peptide bond has important consequences for protein structure.

The effect that this rigidity has on possible protein shapes is to limit somewhat the number of possibilities, and as such, it introduces an element of control. Picture a heavy chain with rather long links. Between the links there is free rotation, but the links themselves are rigid. Compared to a rope, there are less possible ways that the chain could be contorted. However, for the biological functions that proteins perform, having specific ways (rather than infinite ways) that the protein can shape itself gives more control and more specificity to the protein function.

WRITING PEPTIDE AND PROTEIN SEQUENCES: N-TERMINUS TO C-TERMINUS

Peptide and protein sequences are written from left to right as amino, or N-terminus, to carboxyl, or C-terminus. The amino/carboxyl refers to free amino or free carboxyl groups, meaning that they are not part of a peptide bond.

Example:

Draw the dipeptides Phe-Asp and Asp-Phe. They're different! Not only do they have different structures, but their pI values will also be different. Calculate their pI values.

Solution:

Phe–Asp	**Asp-Phe**
^+H_3N–Phe–Asp–COO$^-$	^+H_3N–Asp–Phe–COO$^-$

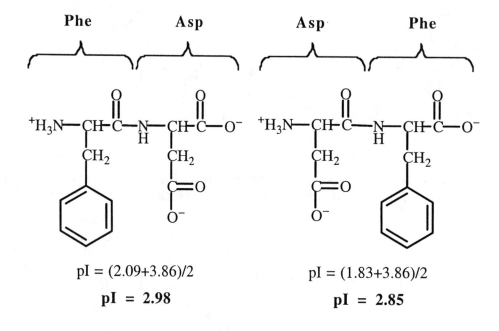

Phe **Asp**	**Asp** **Phe**
pI = (2.09+3.86)/2	pI = (1.83+3.86)/2
pI = 2.98	**pI = 2.85**

Aspartame™

Asp-Phe is an important commercial dipeptide. Its methyl ester (on the C-terminus) is Aspartame™, an artificial sweetener that is 200 times sweeter than sugar. Both amino acids have the L-configuration around the α-carbon. If either is in the D-configuration, then the peptide is bitter rather than sweet.

AMINO ACID SEQUENCING TECHNIQUES

- Carboxypeptidase: C-terminus

- Sanger's method: N-terminus: DNFB reacts with the N-terminal amino acid to form a DNP derivative

- Trypsin: hydrolyzes the C-side of Lys and Arg

- Chymotrypsin: hydrolyzes the C-side of Phe, Tyr, and Trp

- Cyanogen bromide: hydrolyzes the C-side of Met

- Edman degradation: sequentially determines the N-terminal residue of each fragment

THE IMPORTANCE OF ALLOSTERY

Allostery pulls together the complexity and importance of protein 3-D structure and ties them directly to function. It all boils down to those intra- and intermolecular forces (as well as covalent bonds) that determine a protein's overall shape. The shape affects (some might say determines) the protein's function. The idea that a small molecule (or ion) binds to a protein and induces a change in the shape - a conformational change. When the shape changes, the protein's affinity for other ligands also changes. We'll see this in chapter 6, as allosteric enzymes are regulated by the binding of ligands.

That small molecule that binds to a protein is called a ligand, ligand-induced conformational changes are called allosteric transitions and the ligands that trigger them are called effectors or modulators.

AFTER STUDYING THIS CHAPTER, YOU SHOULD BE ABLE TO SOLVE THESE TYPES OF PROBLEMS:

- Classify amino acids as polar (neutral), nonpolar, acidic, or basic

- Draw the structure and give the net charge of an amino acid or peptide at a specific pH. Will it move towards the anode or the cathode in gel electrophoresis?

- Draw (or use) titration curves; determine (or use) pK_a and pI values.

- Name and draw peptides, given the three-letter abbreviations. Given a structure of a peptide, write it out using the three-letter abbreviations.

- Calculate the pI values of peptides.

- Determine the amino acid sequence of a peptide.

- Predict the secondary structure of a given polypeptide: which amino acids tend to stabilize or disrupt each type of secondary structure?

PRACTICE PROBLEMS TO PROMOTE PERFECTION

Answers are on page 57!

1. Given the peptide: Val-Arg-Ala-Tyr-Gly

    a. Draw the structure of the peptide in its most acidic form.
    b. Name it.
    c. Would you expect this peptide to have a high or a low pI? Calculate the pI. (Refer to Table 5.2 in your text for pK_a values.)
    d. Draw the titration curve for this peptide.
    e. What is the net charge of this peptide at pH = 7? In gel electrophoresis, would it move towards the positive anode or the negative cathode?
    f. If this peptide was produced as a fragment during the determination of the amino acid sequence of a protein, which of the following methods could have been used to produce it? Does this give you any further information regarding the primary structure? Methods: carboxypeptidase, Sanger's method (DNFB), trypsin, chymotrypsin, Edman degradation

2. Given the peptide: Pro-His-Met-Ser-Phe

    a. Draw the structure of the peptide in its most acidic form.

    b. Calculate the pI. (Refer to Table 5.2 in your text for pK_a values.)

    c. What is the net charge of this peptide at pH = 12? In gel electrophoresis, would it move towards the positive anode or the negative cathode?

    d. Answer (1f), above, for this peptide.

3. Given the following peptide, write out the proper sequence of amino acids using their three-letter abbreviations.

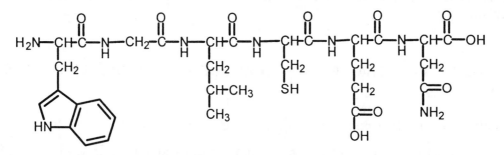

4. Determine the amino acid sequence of a polypeptide given the following data.

    Treatment with carboxypeptidase liberates Val.

    Treatment with DNFB liberates DNP-Ser.

    Treatment with trypsin results in the following fragments:

    Glu-His-Phe-Arg, Pro-Val, Ser-Tyr-Ser-Lys, Val-Trp-Gly-Lys.

    Treatment with chymotrypsin results in the following fragments:

    Arg-Val-Trp, Gly-Lys-Pro-Val, Ser-Lys-Glu-His-Phe, Ser-Tyr.

    Total hydrolysis produces the following amino acids:

    Arg, Glu, Gly, His, Lys(2), Phe, Pro, Ser(2), Trp, Tyr, Val(2).

5. Determine the amino acid sequence of a polypeptide given the following data.

    Treatment with carboxypeptidase liberates Val.

    Treatment with DNFB liberates DNP-Ser.

    Treatment with trypsin results in the following fragments:

    Phe-Glu-His-Lys, Phe-Gly-Arg, Pro-Val, Ser-Tyr-Ser-Lys, Trp-Gly-Lys.

    Treatment with chymotrypsin results in the following fragments:

    Gly-Arg-Trp, Gly-Lys-Pro-Val, Ser-Lys-Phe, Glu-His-Lys-Phe, Ser-Tyr.

    Total hydrolysis produces the following amino acids:

    Arg, Glu, Gly(2), His, Lys(3), Phe(2), Pro, Ser(2), Trp, Tyr, Val.

    (Hint: If you get stuck working from the C-terminal end, try working from the N-terminal end, and vice-versa.)

OPIOID - a great way to get rid of vowels while playing Scrabble®

CHAPTER 5: ANSWERS TO EVEN-NUMBERED REVIEW QUESTIONS

2. a. nonpolar   b. polar   c. acidic   d. basic   e. nonpolar

    f. basic   g. nonpolar   h. polar   i. nonpolar   j. nonpolar.

4. First, draw His at the lowest (acidic) pH. Then, remove one H+ at a time (in order of pK_a) and draw each structure. Remember (when you learned about buffers) that at each of the plateaus in the titration curve, pH = pK_a, and [HA] = [A–]. The pK_a values, then, are equal to the pH at each of the plateaus. It helps to go ahead and label the reaction arrows with the pK_a values and each structure with its net charge. Going through this exercise helps to visualize the rest of the problem.

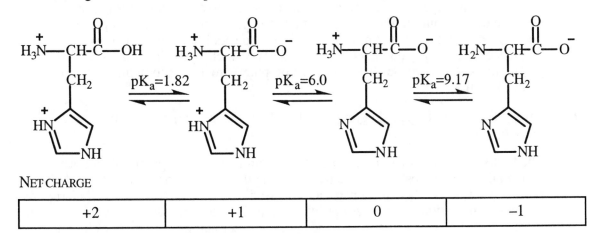

NET CHARGE

+2	+1	0	−1

    a. So, at the first plateau, the pH = 1.82 and the first two species above are present. At the second plateau, the pH = 6.0, and the second and third species are present. At the third plateau, the pH = 9.17, and the third and fourth species are present.

    b. The pKa values for each species are approximately 1.8, 6 and 9.2, respectively.

    c. The isoelectric point for histidine is (6.0 + 9.2)/2 = 7.6

6. The resonance forms of the peptide bond in glycylglycine are as follows:

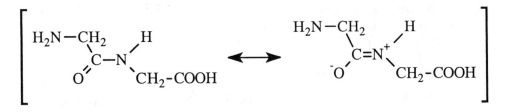

    The partial double bond character of the peptide bond makes it rigid and planar. Rotation around this bond is therefore hindered.

8. a. Fibrous proteins, which possess water-insoluble sheetlike or ropelike shapes, typically have structural roles in living organisms. Globular proteins are compact spherical molecules (usually water-soluble) that typically have dynamic functions.

    b. Simple proteins contain only amino acids. A conjugated protein is a simple protein combined with a nonprotein component, such as lipids or sugars.

c. An apoprotein is a protein without its prosthetic group. An apoprotein molecule combined with its prosthetic group is a holoprotein.

10. a. The amino acid sequence is a polypeptide's primary structure.

   b. β-pleated sheet is one type of secondary structure.

   c. Inter- and intra-chain hydrogen bonds between N-H groups and carbonyl groups of peptide bonds are the principal feature of secondary structure. Hydrogen bonds formed between polar side chains are important in tertiary and quaternary structure.

   d. Disulfide bonds are strong covalent bonds that contribute to tertiary and quaternary structure.

12. The structural features of several amino acids do not foster α-helix formation. Because the R group of glycine is too small, the polypeptide chain becomes too flexible. Proline's rigid ring prevents the required rotation of the N-C bond. Sequences with larger numbers of amino acids with charged side chains (e.g., glutamate) and bulky side chains (e.g., tryptophan) are also incompatible with α-helix formation.

14. Amino acids with basic side groups such as lysine, arginine, or tyrosine would contribute to a high pI value.

16. The first step in the isolation of a specific protein is the development of an assay which allows the investigator to detect it during the purification protocol. Next, the protein, as well as other substances, are released from source tissue by cell disruption and homogenization. Preliminary purification techniques include salting out in which large amounts of salt are used to induce protein precipitation, and dialysis in which salts and other low molecular weight material are removed. Further purification methods, which are adapted to each research effort at the discretion of the investigator, include various types of chromatography and electrophoresis. Three chromatographic methods are: ion-exchange chromatography, gel-filtration chromatography, and affinity chromatography. Gel electrophoresis may be used to purify a protein and/or to assess the purity of a protein.

18. In sequencing a polypeptide, the next residue in a sequence is determined by the increase in height of the peak on the amino acid analyzer. If there is already a large amount of that amino acid present in the solution, it would be difficult, if not impossible, to detect any change in peak height. Small fragments have only a few amino acids, and this problem does not occur.

20. The structure of β-endorphin is

   Tyr–Gly–Gly–Phe–Met–Thr–Ser–Glu–Lys–Ser–Gln–Thr–Pro–Leu–Val–Thr–Leu–

   Phe–Lys–Asn–Ala–Ile–Val–Lys–Asn–Ala–His–Lys–Lys–Gly–Gln

22. The following fragments are produced when bradykinin is treated with the indicated reagents:

   a. carboxypeptidase: Arg and Arg-Pro-Pro-Gly-Phe-Ser-Pro-Phe

   b. chymotrypsin: Arg-Pro-Pro-Gly-Phe, Ser-Pro-Phe, and Arg

   c. trypsin: Arg and Pro-Pro-Gly-Phe-Ser-Pro-Phe-Arg

   d. DNFB: DNP-Arg and the following amino acid residues:

          3 Pro, 1 Gly, 2 Phe, 1 Ser, and 1 Arg

CHAPTER 5: ANSWERS TO EVEN-NUMBERED THOUGHT QUESTIONS

2. Living cells possess complex mechanisms for assisting the proper folding of nascent polypeptides. These mechanisms are poorly understood and cannot yet be duplicated in the laboratory.

4. The immobilized water of protein molecules is locked into position by hydrogen bonding between polar and ionic groups and water molecules. This gives rise to a three-dimensional structure in which the water molecules have very little freedom of motion, i.e., they are frozen in place.

6. The hydrophobic amino acid side chains are excluded from the water and tend to cluster together. This clustering holds portions of the polypeptides in a particular conformation.

8. Proline and hydroxyproline are both imino acids and do not lose their nitrogen atom when they react with ninhydrin. As a result, when proline reacts with ninhydrin, the compound shown below is formed.

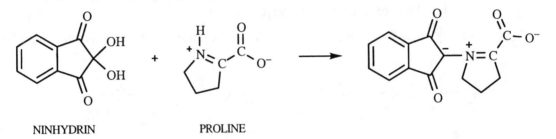

NINHYDRIN   PROLINE

ANSWERS TO "PRACTICE PROBLEMS TO PROMOTE PERFECTION"

1. a. The pK_a value for each ionizable group is shown in bold.

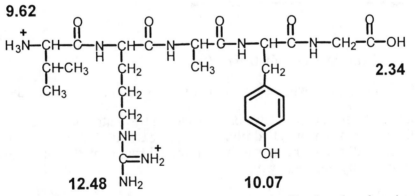

b. valylarginylalanyltyrosylglycine (Remove the "ine" and replace it with "yl.")

c. We'd expect this peptide to have a high pI, since the side groups are either basic or neutral. To calculate the pI, determine the structure that has a net charge of zero, and average the pK_a values on either side of that structure. The best way to do this is to draw each form in order of pK_a, that is, as the peptide is titrated.

The pI is the average of 9.62 and 10.07, or $(9.62+10.07)/2 = 9.85$. (Note that if this were a multiple-choice problem, one wrong choice would probably be 8.63, the average of *all* of the pK_a values.)

d. To draw the titration curve, first label the y-axis "pH" and the x-axis "Equivalents OH⁻." Draw short plateaus at each pK_a value. Place a point at a pH value that is midway between each two pK_a values. Draw a smooth curve from the plateaus through each inflection point.

e. At pH = 7, the peptide has a +1 charge, and would move towards the cathode.

f. Val-Arg-Ala-Tyr-Gly

Since we have a fragment and not a single amino acid, we can rule out the N-terminus or the C-terminus determination methods. We can also rule out chymotrypsin, since the Tyr-Gly peptide bond remains intact, and trypsin, since the Arg-Ala bond remains intact. That leaves cyanogen bromide. The amino acid that is linked to the N-terminal side of Val must be Met, the amino acid whose C-side peptide bond is hydrolyzed by cyanogen bromide. Since we do not have a method that hydrolyzes at the C-terminal side of Gly, this fragment must be located at the C-terminal end of the protein.

2. a. The pK_a value for each ionizable group is shown in bold. Pro-His-Met-Ser-Phe:

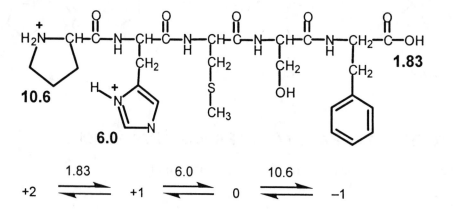

b. The pI is 8.3 (the average of 6.0 and 10.6).

c. At pH=12, the peptide will have a −1 charge, and will move towards the anode.

d. This peptide fragment could have been produced by treatment with chymotrypsin, since the C-side of Phe was hydrolyzed. If either Lys or Arg was linked to the N-side of Pro, and this fragment was the C-terminal fragment, then it could have been produced by treatment with trypsin.

3. Trp-Gly-Leu-Cys-Glu-Asn

4. Ser-Tyr-Ser-Lys-Glu-His-Phe-Arg-Val-Trp-Gly-Lys-Pro-Val

5. Ser-Tyr-Ser-Lys-Phe-Glu-His-Lys-Phe-Gly-Arg-Trp-Gly-Lys-Pro-Val

CHAPTER SIX

Enzymes

PROPERTIES OF ENZYMES

Activation energy (E_a, or free energy of activation), substrate, transition state

Enzymes catalyze a wide variety of chemical reactions that would otherwise take place at impossibly slow rates. Enzymes accelerate the rate of reaction by stabilizing the transition state of a reaction. This stabilization lowers the activation energy, which is the energy a substrate must have before it can be transformed into product. In other words, the activation energy of the enzyme-substrate complex is lower than that of the substrate alone.

Enzyme advantages over chemical catalysts:

1. Enzyme-catalyzed reactions often have *much* higher rates.

2. Enzymes have greater reaction specificity.

3. Enzymes can be regulated.

Enzymes don't affect the thermodynamics (e.g., the ΔG and the equilibrium constant) of a reaction. If a chemical reaction isn't thermodynamically favorable ($+\Delta G$), an enzyme can't change this. Enzymes help spontaneous reactions to achieve equilibrium *faster*. Without enzymes, essential reactions would take place too slowly to allow life to be possible!

Enzymes are biological catalysts - at the end of a reaction, they're intact, back to their original form, and ready to catalyze another reaction. That's why we typically don't need high enzyme concentrations, and why all enzymes act at nanomolar or picomolar concentrations under physiological conditions.

Enzyme-catalyzed reactions are typically highly specific to a particular substrate. For example, a particular enzyme might catalyze a reaction with L-alanine but not D-alanine or any other amino acid. The substrate binds to the enzyme at the active site by non-covalent interactions, including hydrogen bonding, electrostatic, and hydrophobic forces. The active site of an enzyme provides the specificity for determining which chemical reactions will be catalyzed by that enzyme. Contact points at the active site have a particular orientation and a distinct shape, like a right-handed glove. Only the right hand (not a foot or even a left hand!) will fit.

Lock-and-Key Model vs. Induced Fit Model

The lock and key model for enzyme action is easy to visualize: the substrate fits like a key into the active site lock. But think about it. If it fits so perfectly, why would it want to change, or react, at all? Hence the induced fit model. Picture this: the substrate fits o.k. but not perfectly. Its presence in the active site causes a conformational change in the enzyme that encourages the transition state, and lowers the energy of the transition state. If the substrate can get to the transition state faster, it's all downhill from there![1]

---

[1] Of course this is oversimplified, and we have more in-depth models that explain enzyme action more thoroughly. In particular, see concerted vs. sequential mechanisms of enzyme action.

CLASSIFICATION OF ENZYMES ACCORDING TO TYPE OF REACTION CATALYZED

Oxidoreductases	Hydrolases	Isomerases
Transferases	Lyases	Ligases

Examples of Each Type of Enzyme Class[2]

★ OXIDOREDUCTASE

Reaction Type:	Oxidation-reduction
Names Include:	dehydrogenase, oxidase, oxygenase, reductase, peroxidase, hydroxylase
Example:	Alcohol dehydrogenase
Function:	Catalyzes the transfer of electrons from an alcohol to NAD$^+$, producing the corresponding aldehyde. This is the first step in the catabolism of alcohols. Electrons are removed from the alcohol carbon to form the aldehyde.

$$CH_3CH_2OH + NAD^+ \rightleftharpoons CH_3-\overset{\overset{O}{\|}}{C}-H + NADH + H^+$$

ETHANOL   ACETALDEHYDE

★ TRANSFERASE

Reaction Type:	Transfer of functional groups
Names Include:	kinase (phosphate group transfer), transferase
	Also: "trans" + (group transferred) + "ase"
	Examples: transcarboxylase, transmethylase, transaminase
Example:	Catechol N-methyltransferase
Function:	Catalyzes the transfer of a methyl group from S-adenosyl-methionine to norepinephrine. Used in the synthesis of epinephrine, a neurotransmitter.

NOREPINEPHRINE   EPINEPHRINE

★ HYDROLASE

Reaction Type:	Hydrolysis (cleave bonds by adding water)
Names Include:	esterase, phosphatase, peptidase, protease

---

[2] Further examples may be found in Table 6.1, page 165 of your text.

Serine proteases use the –CH$_2$OH side chain of serine to hydrolyze a peptide. Examples of serine proteases: trypsin, chymotrypsin, thrombin

Example:	Thrombin
Function:	Hydrolyzes the peptide bond after an arginine. Converts fibrinogen into fibrin; fibrin then polymerizes to form a blood clot.

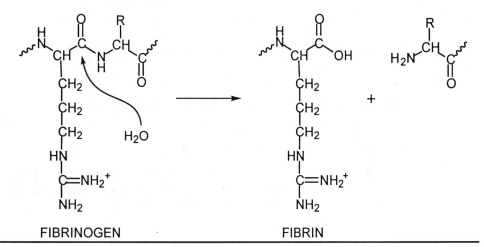

FIBRINOGEN   FIBRIN

★ LYASE

Reaction Type:	Removes a group and forms a double bond (or the reverse reaction: Adds a group to a double bond)
Names Include:	decarboxylase, hydratase, dehydratase, deaminase, synthase
Example:	Argininosuccinate lyase
Function:	Converts argininosuccinate to arginine and fumarate in urea synthesis. Removes a four-carbon dicarboxylic acid group, forming a double bond between the two central carbons.

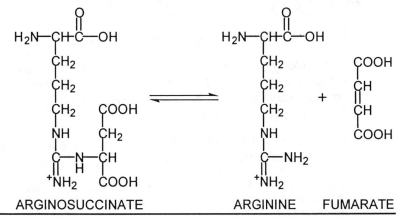

ARGINOSUCCINATE   ARGININE   FUMARATE

★ ISOMERASE

Reaction Type:	Isomerization
Names Include:	mutase (transfer of functional group(s) within a molecule) epimerase (inversion of stereochemistry at one chiral carbon)
Example:	Triose phosphate isomerase

Function: Interconverts glyceraldehyde-3-phosphate and dihydroxyacetone phosphate. The carbonyl group is moved from
carbon 1 in glyceraldehyde to carbon 2 in dihydroxyacetone.
Involved in glycolysis or sugar fermentation.

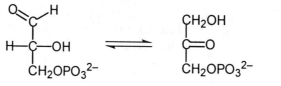

Glyceraldehyde-3-phosphate   Dihydroxyacetone phosphate

✦ LIGASE

Reaction Type: Bond formation with ATP hydrolysis

Names Include: synthetase, carboxylase

Example: Glutamine synthetase

Function: Catalyzes the formation of glutamine from glutamate and
ammonia using ATP hydrolysis to make the reaction
thermodynamically favorable.

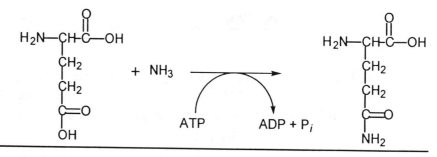

ENZYME KINETICS

Reaction order

FIRST-ORDER AND SECOND-ORDER KINETICS:

The velocity (v), or rate, of any chemical reaction is the change in concentration of substrate (or product) as a function of time. The rate is proportional to the substrate concentration raised to the power n, where n is the order of the reaction (and k is the rate constant): $v = k[S]^n$

For the simple first order reaction $S \rightarrow P$, the rate $= v = k[S]$. (If $[S]$ is doubled, for example, v should also double.)

The order of a reaction depends on the reaction mechanism, specifically, how many reactants (or substrates) are involved in the slowest step of the mechanism (because the rate can't go faster than the slowest step). For example, consider the reaction:

$A + B \rightarrow C + D$. The general rate equation is $v = k[A]^m[B]^n$ (where m is the order with respect to A, and n is the order with respect to B). If the rate only depends upon the concentration of A, then $v = k[A]^1[B]^0 = k[A]$, and the reaction is first order overall. If the rate depends on the concentrations of both A and B, then $v = k[A][B]$, and the reaction is second order overall. It's important to keep in mind that the order of a reaction must be determined experimentally. There's no way to know the order of a reaction just by looking at the chemical equation.

Pseudo-first-order kinetics is a second-order reaction that behaves as if it's first order, typically because one reactant (such as water) is present in excess.

ZERO-ORDER KINETICS: n = 0

With zero-order kinetics, the reaction rate doesn't depend on [S]. Adding more substrate won't increase the rate. Why not? If all of the active sites are filled (saturated) with substrate (and there's plenty of substrate around to hit any active site that becomes available), then adding more substrate won't increase the rate. The enzyme can't work any faster – it's at its maximum velocity (V_{max}).

⭐ Michaelis-Menten Kinetics

Michaelis-Menten equation:	*Michaelis-Menten plot:*
$$v_0 = \frac{V_{max}[S]}{[S] + K_m}$$	v_0 vs. [S]

V_{max} = maximum velocity

K_m = Michaelis constant (experimentally determined) = [S] at half of V_{max}

When [S] = K_m, then the Michaelis-Menten equation becomes $v_0 = \dfrac{V_{max}[S]}{2[S]}$,

which simplifies to $v_0 = \dfrac{V_{max}}{2}$. In other words, **the K_m is equal to the substrate concentration when v_0 equals half of V_{max}.** So, K_m can be determined from the Michaelis-Menten plot of v_0 vs. [S].

Keep in mind that the Michaelis-Menten plot is actually a *series* of experiments that measure the *initial rate* of a reaction at a number of different substrate concentrations, with the same enzyme concentration in each experiment. Take a second look at Figure 6.3 on page 168. Figure (b) shows one experiment. Figure (a) would be data from a number of experiments like (b), each point representing an experiment that begins with a different [S].

⭐ ASSUMPTIONS MADE TO DERIVE THE MICHAELIS-MENTEN EQUATION:

1. Steady-state approximation: The enzyme-substrate complex (ES) is in steady state, which means the concentration of ES remains constant as a function of time. ES is formed at the same rate that it's removed.

2. At saturation, all of the enzyme is in the ES form (when the [S] >>>[E], there's no free enzyme).

3. When all the enzyme is in the ES form, the reaction rate is at its maximum.

k_{cat} = turnover number = # substrate molecules converted to product per second;

   k_{cat}/ K_m is a gauge of catalytic efficiency

enzyme efficiency =

Diffusion control limit - the rate is limited by the rate of diffusion. A reaction can't go faster than it takes for the substrate to get to the active site.

Catalytic perfection

International unit (I.U.) of enzyme activity = amount of enzyme to produce 1μmol of product per minute; Specific activity = # I.U. per mg of protein; a measure of enzyme purification

katal (kat) = 1 mole of substrate converted to product per second

★ K_M AS AN INDICATION OF AN ENZYME'S AFFINITY FOR SUBSTRATE

enzyme affinity =

The *lower* the K_m (Michaelis constant), the *greater* the enzyme's affinity for the substrate. Why is this true? Consider two enzyme-catalyzed reactions with the same V_{max} but different K_m values. For the reaction in which the enzyme has a greater affinity for substrate, it takes less substrate to get to the same rate (half of V_{max}). Conversely, an enzyme with a lower affinity for substrate needs a higher substrate concentration to work at the same velocity.

★ Lineweaver-Burk Plots: Determination of K_m and V_{max}

SIGNIFICANCE:

- Since a Lineweaver-Burk plot is a straight line, the values of V_{max} and K_m are much easier to determine than by using a Michaelis-Menten plot, in which the velocity approaches (but never reaches) a maximum value.

- Comparing Lineweaver-Burk plots of inhibited vs. uninhibited reactions can indicate the mechanism of enzyme inhibition that is occurring.

Lineweaver-Burk turned the Michaelis-Menten plot into a straight line by taking the reciprocal of the Michaelis-Menten equation ($v_0 = \dfrac{V_{max}[S]}{[S] + K_m}$) and rearranging it to fit the equation for a straight line, $y = mx + b$. Doing further algebra shows that the x-intercept = $-1/K_m$.

$$\frac{1}{v_0} = \frac{[S] + K_m}{V_{max}[S]} = \frac{[S]}{V_{max}[S]} + \frac{K_m}{V_{max}[S]} = \frac{1}{V_{max}} + \frac{K_m}{V_{max}[S]}$$

Rearranged slightly, this equation becomes:	
$\dfrac{1}{v_0} = \dfrac{K_m}{V_{max}} \bullet \dfrac{1}{[S]} + \dfrac{1}{V_{max}}$	$m =$ slope $= \dfrac{K_m}{V_{max}}$
$y \quad = \quad m \quad x \quad + \quad b$	$b =$ y-intercept $= \dfrac{1}{V_{max}}$
A plot of $\dfrac{1}{v_0}$ vs. $\dfrac{1}{[S]}$ gives a straight line.	x-intercept $= -\dfrac{1}{K_m}$

So, to calculate K_m and V_{max}, all that's needed is the x and y intercepts from the Lineweaver-Burk plot.[3] To keep these straight, remember that the x-axis is a measure of substrate concentration, from which we get K_m, and that the y-axis is a measure of velocity.

Note: A Lineweaver-Burk plot is sometimes referred to as a *double-reciprocal* plot, since it's $\dfrac{1}{v_0}$ vs. $\dfrac{1}{[S]}$ instead of the Michaelis-Menten plot of v_0 vs. [S].

Enzyme Inhibition and Inhibitors

- Inhibitors bind to an enzyme and interfere with its activity. Inhibitors help to regulate enzyme activity, essentially regulating entire pathways as a result.

- Types of reversible inhibition: competitive, uncompetitive, noncompetitive (pure and mixed)

- To determine the type of inhibitor, use Lineweaver-Burk plots of inhibited and uninhibited reactions.

- Allosteric enzymes (Michaelis-Menten plot is sigmoidal rather than hyperbolic)

    Effectors bind to allosteric or regulatory sites and change enzyme activity.

    Most are multisubunit enzymes, and are located at key regulatory steps in biochemical pathways.

- Irreversible inhibitors bind to an enzyme covalently and kills all enzyme activity.

MECHANISMS FOR REVERSIBLE ENZYME INHIBITION:

1. Competitive: $E + I \Leftrightarrow EI$ $E + S \Leftrightarrow ES \rightarrow P$

    Competitive inhibitors bind to the active site of the free enzyme; they "compete" or prevent the enzyme from binding substrate. This results in a change in K_m (since it affects substrate binding) but not V_{max} (since the effect of the inhibitor can be overcome by increasing [S]). In other words, it takes much more substrate to be able to reach V_{max}.

2. Uncompetitive: $ES + I \Leftrightarrow ESI$ $E + S \Leftrightarrow ES \rightarrow P$

    Uncompetitive inhibitors bind to ES rather than to E, the free enzyme. Since some of the enzyme is always tied up as ESI, V_{max} can't be attained, and is lower. K_m also *decreases*. How can an inhibitor *increase* the affinity of an enzyme for its substrate, and still inhibit the reaction? Think on this: If the inhibitor binds to ES and removes ES, then LeChatelier's principle says that the equilibrium will shift to compensate for the change in concentration. The reaction ($E + S \Leftrightarrow ES$) is shifted to the right because the inhibitor binds to and removes ES. So, the enzyme binds more substrate than expected (decreasing K_m), but the ES complex is waylaid by inhibitor before it can be

---

[3] Note that the x-intercept is just a mathematical trick to calculate K_m, since a negative substrate concentration has no real physical meaning. The calculation of V_{max} is also a mathematical trick, since reaction rate at [S]=0 also has no physical meaning.

turned into product (decreasing V_{max}). Adding more [S] can't overcome this – even though more ES would form, more ESI would also form.

3. Noncompetitive (pure, mixed):

$$E+I \Leftrightarrow EI \qquad\qquad ES + I \Leftrightarrow ESI \qquad\qquad E + S \Leftrightarrow ES \rightarrow P$$

Noncompetitive inhibitors bind to a site other than the substrate binding site; they can remove both free enzyme (E) and ES. In pure noncompetitive inhibition (the simplest, but less common, case), noncompetitive inhibitors change the V_{max} but not the K_m. Noncompetitive inhibition can't be reversed by increasing [S]. Mixed inhibition changes both K_m and V_{max}.

SUMMARY OF REVERSIBLE ENZYME INHIBITION

Type:	*Binds to:*	*Affects:*	*L-B Plot:[4]*	*Notes:*
Competitive	active site of E, not ES	higher K_m same V_{max}	lines intersect on *y*-axis	I is often similar to substrate's structure
Uncompetitive	ES, not E	lower K_m lower V_{max},	slopes are the same, *x*- and *y*-intercepts differ	typically for reactions with more than one S
Noncompetitive PURE	both E and ES (at a site other than the active site)	same K_m lower V_{max}	lines intersect on *x*-axis	rare; the rate to form EI equals the rate to form ESI
Noncompetitive MIXED		changes K_m lower V_{max}	*x*- and *y*-intercepts and slopes differ	the rate to form EI doesn't equal the rate to form ESI

CATALYSIS

Mechanisms

Proximity and Strain Effects; Electrostatic Effects

Basically, the more tightly and efficiently that an active site can bind substrate while it's in its transition state, the faster the reaction rate will be.

Acid-Base Catalysis

Covalent Catalysis (example: serine proteases like chymotrypsin)

Cofactors: structural properties and chemical reactivities

Cofactors are non-protein components that add chemical functionality to enzymes, and thus increase the kinds of reactions that enzymes can catalyze. (Otherwise, the enzymes would be limited to the amino acid residues' R groups.)

---

[4] Lineweaver-Burk plot of inhibited and uninhibited reactions: Increasing Km makes the *x*-intercept $(-1/Km)$ less negative (-1/20 is less negative than $-1/10$). Lowering $Vmax$ makes the *y*-intercept($1/Vmax$) larger (1/2 is larger than 1/4). The slope is $Km/Vmax$.

APOENZYME VS. HOLOENZYME

In *A*poenzymes, the necessary cofactor or coenzyme is *A*bsent. (An apoenzyme is just the protein component.) *H*oloenzymes *H*ave the cofactor or coenzyme.

METALS: transition metals (Fe^{2+}, Cu^{2+})

alkali and alkaline earth metals (Na^+, K^+, Mg^{2+}, Ca^{2+})

Why are the transition metals good cofactors? Their high concentration of positive charge binds small molecules well. They can act as Lewis acids (i.e., accept electron pairs). Because they can interact with two or more ligands, they can form a substrate-metal ion complex in the active site, thus helping to orient the substrate properly, polarizing the substrate and promoting catalysis. Metals with two or more valence states can mediate redox reactions.

COENZYMES VS. VITAMINS

Vitamins: water-soluble vs. lipid-soluble; serve as precursors to coenzymes

The vitamin niacin is a precursor to the coenzyme NAD^+ (nicotinamide adenine dinucleotide), which functions as an intracellular electron carrier and as a hydride ($H{:}^-$) transfer agent.

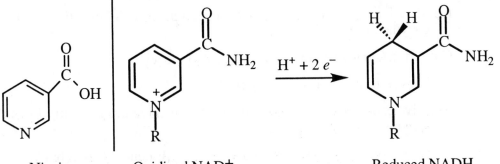

| Niacin | Oxidized NAD$^+$ | Reduced NADH |

The vitamin riboflavin is a precursor to the coenzymes FMN (flavin mononucleotide) and FAD (flavin adenine dinucleotide), which are components of flavoproteins (a sub-class of the oxidoreductases).

Effects of temperature and pH on enzyme-catalyzed reactions

Optimum temperature depends on pH and ionic strength.

higher temperature = faster rate, but too high = denaturation of enzyme

pH optimum (varies according to enzyme function)

Detailed Mechanisms of Enzyme Catalysis: Chymotrypsin and Alcohol Dehydrogenase

Note that these mechanisms resemble those that you wrote in organic chemistry, but the proximity of certain functional groups to each other and the geometry of the active site encourages reactions to happen that would not normally happen. It's like one of those romantic comedies: two people were meant to be together (their being together is thermodynamically favorable) but they'd never fall in love if they hadn't had something put them together, like being stuck in an elevator (or in an enzyme's active site). Again, enzymes can't make the impossible happen – they just make spontaneous reactions happen *faster*.

✶ ENZYME REGULATION

Enzyme regulation maintains an ordered state, conserves energy, and helps the cell to respond to environmental changes (i.e., modulates specific pathways in response to a cell's needs). Regulatory enzymes in a pathway are usually controlled by covalent modification or allosteric regulation.

Covalent Modification

- Phosphorylation (or dephosphorylation) to convert an enzyme between its active and inactive forms (Whether the phosphorylated enzyme is active or inactive depends on the specific enzyme.)

- Zymogens (proenzymes) are converted into active enzymes by the irreversible cleavage of one or more peptide bonds.

Allosteric Regulation

Effector molecules, protomers

Homotropic allostery (cooperativity) vs. heterotropic effects

Negative feedback inhibition - a product of a pathway inhibits the activity of an enzyme early in the pathway

Concerted (symmetry) model vs. sequential model

T vs. R forms (taut vs. relaxed)

Negative cooperativity - binding of the first ligand reduces the affinity of the enzyme for similar ligands

Positive cooperativity - first ligand increases subsequent ligand binding

Importance of the flexibility of proteins (binding of ligand to one protomer prompts a conformational change that's transmitted to other protomers)

Genetic Control

Enzyme induction: Enzymes are synthesized when needed.

Repression: Enzyme synthesis is inhibited.

Compartmentation

In compartmentation, enzymes, substrates and regulatory molecules are physically:

- separated into different regions or compartments to use resources efficiently (e.g., to prevent a just-synthesized molecule from being degraded)

- located close together to help the efficiency of coupled reactions or reactions of a particular pathway (such as the electron transport chain)

Eukaryotes exhibit compartmentation via organelles, that is, some pathways take place exclusively inside (or outside) a particular organelle. That gives the cell an additional level of regulation by controlling the transport of substrates, products, effectors, etc. across organelle membranes (example: mitochondria membranes). Also, special microenvironments (e.g., high pH) can be created within organelles.

✦THE ORGANIC CONNECTION: NICOTINAMIDE ADENINE DINUCLEOTIDE (NAD)

Take another look at the nicotinamide rings of NAD^+ and NADH. Which one would you expect to be more stable, that is, at a lower energy? Why?

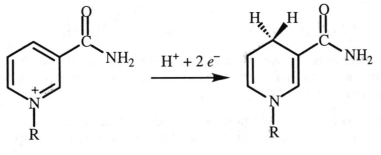

Oxidized NAD^+ Reduced NADH

When the hydride (or, the H^+ with two electrons) is added, it creates a tetrahedral carbon that interrupts the aromaticity of the ring. The nicotinamide ring of NAD^+ is aromatic and thus more stable and lower-energy than the non-aromatic ring of NADH. Since NADH is at a higher energy than NAD^+, this half-reaction would require an input of energy:

$$NAD^+ + H^+ + 2e^- \rightarrow NADH$$

Couple this with an oxidation half-reaction that releases energy, and we have a neat way of capturing that energy in the form of electrons. (Sound familiar? Refer to Chapter 4.)

AFTER STUDYING THIS CHAPTER, YOU SHOULD BE ABLE TO SOLVE THESE TYPES OF PROBLEMS: *(R = Review Questions, T = Thought Questions)*

- Determine the class of an enzyme, given the reaction that the enzyme catalyzes. (R4)

- Given a biochemical reaction and reaction rate data at various substrate concentrations, determine the order of a reaction for each substrate and the overall order of a reaction. (T1)

- Michaelis-Menten plots: Determine K_m. (T2)

- Lineweaver-Burk plots: Determine the type of enzyme inhibition, given data for both uninhibited and inhibited enzyme-catalyzed reactions. Determine V_{max} and K_m from the *x*- and *y*-intercepts. (T3, T5, T9)

- Compare enzyme efficiencies given k_{cat} and K_m data for various substrates. (T8)

APPLICATION OF ENZYME KINETICS

Some human populations, such as Native Americans and Asians, may be more sensitive to alcoholic beverages than others. The consumption of small amounts of ethanol in some individuals produces vasodilatation, which results in facial flushing or turning red. This physiological response to alcohol arises from acetaldehyde, which is generated by liver alcohol dehydrogenase.

$$CH_3CH_2OH + NAD^+ \rightarrow CH_3CHO + H^+ + NADH$$

The acetaldehyde that forms is removed by a mitochondrial aldehyde dehydrogenase that converts CH_3CHO into acetate. This enzyme has a low K_m. There also exists a cytosolic aldehyde dehydrogenase with a much higher K_m.

Provide an explanation as to why certain populations may be more sensitive to alcohol than others. How would you test your explanation, and what results would you expect?

Explanation:

Individuals who are hypersensitive to ethanol often lack the mitochondrial form of aldehyde dehydrogenase. As a result, only the low affinity (high K_m) cytosolic enzyme is left to remove acetaldehyde. The concentration of acetaldehyde is thus elevated after alcohol consumption, accounting for increased sensitivity.

To test this explanation, an enzyme assay would be performed in order to measure acetaldehyde dehydrogenase activity in either the cytosol or mitochondria. One would expect to find both the high and low K_m forms of the enzyme in some individuals, and only the high K_m form in individuals sensitive to alcohol.

DRUGS AND ENZYME INHIBITORS

Many useful drugs are enzyme inhibitors. For example, Lovastatin is used to treat high cholesterol levels (hypercholesterolemia). It blocks the synthesis of cholesterol by inhibiting a key enzyme in cholesterol synthesis, HMG-CoA reductase.

The antibiotic penicillin is also an enzyme inhibitor. Penicillin blocks the enzyme that bacteria use to make their cell walls. Since the cell wall is important in protecting the bacteria from environmental changes, preventing its synthesis is detrimental to the bacteria.

Aspirin (salicylic acid) is an enzyme inhibitor that prevents the conversion of arachidonic acid (a membrane lipid) to prostaglandins and thromboxanes by cyclooxygenase.

NIACIN AND PELLAGRA

A diet deficient in the vitamin niacin results in the disease pellagra, characterized by diarrhea, dermatitis, and dementia. Pellagra was a significant health problem around the turn of the century in the rural south where corn was a staple food. Humans normally synthesize niacin from the amino acid tryptophan; corn, however, has very little tryptophan or usable niacin. While corn does have niacin, it is in a form that cannot be absorbed by the intestines. Native Americans, who originally domesticated corn, found a way to convert niacin into a form which could then be absorbed by humans. They accomplished this by first soaking the corn meal in lime water (calcium hydroxide), a process which converts the niacin into a useable form.

CHAPTER 6: ANSWERS TO EVEN-NUMBERED REVIEW QUESTIONS

2. The important properties of enzymes are high catalytic rates, a high degree of substrate specificity, negligible formation of side products and capacity for regulation.

4. (a) oxidoreductase, (b) transferase, (c) lyase, (d) isomerase, (e) ligase.

6. Three reasons that the regulation of biochemical processes are important are maintenance of an ordered state, conservation of energy, and responsiveness to environmental cues.

8. In the concerted model, the substrate and activators bind to the relaxed conformation. This binding shifts the equilibrium to the R conformation. In the sequential model, binding the activator molecule changes the conformation of the enzyme to a shape more favorable to binding substrate. When oxygen binds to hemoglobin, the first oxygen molecule binds slowly. It, however, introduces a conformational change that makes the sequential binding of the second, third, and fourth oxygen molecules much easier.

10. Transition metal ions are useful as enzyme cofactors because they have concentrations of positive charge, can act as Lewis acids, and can bind to two or more ligands at the same time.

12. Enzymes decrease the activation energy required for a chemical reaction because they provide an alternate reaction pathway that requires less energy than the uncatalyzed reaction. They do so principally because of the unique intricately shaped active sites which possess strategically placed amino acid side chains, cofactors, and coenzymes that actively participate in the catalytic process.

14. The pK_a of the imidazole group of histidine is approximately 6. Therefore, the histidine side chain ionizes within the physiological pH range. The protonated form of histidine is a general acid, and the unprotonated form is a general base.

CHAPTER 6: ANSWERS TO EVEN-NUMBERED THOUGHT QUESTIONS

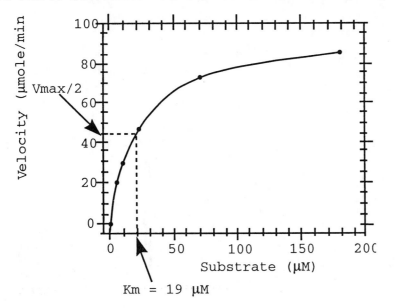

2.

4. Two major types of enzyme inhibitors are competitive and noncompetitive inhibitors. Malonate, whose structure resembles that of succinate, is a competitive inhibitor of succinate dehydrogenase. AMP is a noncompetitive inhibitor of fructose bisphosphate phosphatase.

6. The substrate binds to the active site and helps preserve its structure by acting as a template. In this way, it stabilizes the enzyme in that conformation.

8. The k_{cat}/K_m ($M^{-1}s^{-1}$) values for the substrates are as follows:

Substrate	k_{cat}/K_m
Ethanol	0.5
1-butanol	1.02
1-hexanol	2.63
12-hydroxydodeconate	2.92
all-*trans*-retinol	3.9
benzyl alcohol	0.2
2-butanol	0.00114
cyclohexanol	0.0039

The larger the k_{cat}/K_m value, the faster the rate of the reaction. The substrate with the highest k_{cat}/K_m value is all-*trans*-retinol.

CHAPTER SEVEN

Carbohydrates

FUNCTIONS OF CARBOHYDRATES:

energy sources, structural elements, precursors in the synthesis of biomolecules, biological information storage

MONOSACCHARIDES

Name by type (aldose, ketose), number of carbons (pentose, hexose), or both (aldohexose, ketohexose)

Fischer projections

Simplest sugars: glyceraldehyde, dihydroxyacetone

Monosaccharide Stereoisomers

⭐ Number of possible stereoisomers = 2^n, where n = # of chiral carbon atoms

⭐ D- vs. L- Determine by comparing with D- and L- glyceraldehyde: compare the chiral carbons furthest from the carbonyl carbons

Enantiomers vs. diastereomers vs. epimers

Epimers are diastereomers that differ in configuration at only one chiral center.

Example: Of the following three monosaccharides, which are epimers of each other?   D-glucose, D-allose, D-altrose

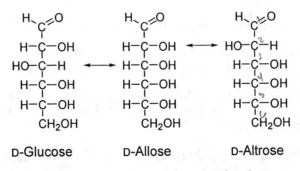

D-Glucose   D-Allose   D-Altrose

D-Glucose and D-allose are epimers of each other because they differ by only one chiral carbon (at carbon 3). D-allose and D-altrose are also epimers of each other since they differ only at carbon 2. However, D-glucose and D-altrose are *not* epimers of each other because they're different at *both* carbons 2 and 3.

Cyclic Structure of Monosaccharides

Formation of cyclic hemiacetals or hemiketals

Anomers (anomeric carbon atom): α vs. β

HAWORTH STRUCTURES OF PYRANOSE AND FURANOSE (CYCLIC) FORMS

Drawing Haworth structures from Fischer projections

CONFORMATIONAL STRUCTURES (CHAIR FORM)

Reactions of Monosaccharides

MUTAROTATION - When α and β monosaccharides are dissolved in water, they interconvert, forming an equilibrium mixture of open-chain, α and β pyranose and furanose forms. The open-chain form can undergo redox reactions.

OXIDATION OF ALDOSES

Types of oxidized aldoses:

aldonic acid: aldehyde to carboxylic acid

uronic acid: terminal $-CH_2OH$ to carboxylic acid

aldaric acid: both aldehyde and terminal $-CH_2OH$ to carboxylic acids

Lactones can be formed from aldonic and uronic acids.

Reducing sugars can be oxidized by Benedict's reagent (a weak oxidizing agent) only if it can exist in its open-chain (aldehyde) form.

REDUCTION of aldehyde and ketone groups to alcohols = alditols

ISOMERIZATION via an enediol intermediate

Aldose-ketose interconversion

Epimerization

ESTERIFICATION: formation of phosphate and sulfate esters

Phosphate esters - important in metabolism; convert $-OH$, a terrible leaving group, into a good leaving group for nucleophilic substitution

Sulfate esters - in proteoglycan components of connective tissue; charged, bind large amounts of water and small ions; form sulfate bridges between carbohydrate chains

GLYCOSIDE FORMATION: cyclic hemiacetal to acetal; cyclic hemiketal to ketal

Glycosidic linkage, glycoside

(fructoside vs. glucoside, furanoside vs. pyranoside)

The glycosidic linkage locks the ring in its cyclic form, so it can't oxidize or mutarotate. (So, β-methyl glucoside is not a reducing sugar.)

Aglycones - noncarbohydrate components of glycosides

Disaccharide, polysaccharide formation

Important Monosaccharides

GLUCOSE - primary fuel for living cells, preferred energy source of brain cells and cells with few or no mitochondria, building block for cellulose, starch, glycogen

FRUCTOSE - a ketohexose isomer of glucose; fruit sugar; used as a sweetening agent (twice as sweet as sucrose)

GALACTOSE - an epimer of glucose; needed to synthesize lactose, glycolipids, certain phospholipids, proteoglycans, glycoproteins; can be made from glucose

In galactosemia, an enzyme needed to metabolize galactose is missing.

Monosaccharide Derivatives

URONIC ACIDS (terminal –CH_2OH group is oxidized to a carboxylic acid)

D-glucuronic acid - combines with molecules to improve their water solubility (and help to remove waste products from the body)

L-iduronic acid = epimer of D-glucuronic acid - both are in components of connective tissues

AMINO SUGARS: AMINO GROUP REPLACES AN –OH GROUP (USUALLY ON CARBON 2)

Common in complex carbohydrate molecules that are attached to cellular proteins and lipids

Most common amino sugars (in animals): D-glucosamine, D-galactosamine

Acetylated amino sugars: N-acetyl-glucosamine, N-acetyl-neuraminic acid, sialic acids

DEOXYSUGARS: –H REPLACES AN –OH

L-Fucose - formed from D-mannose; found in glycoproteins (example: ABO blood group determinates on the surface of red blood cells)

2-Deoxy-D-ribose is the pentose sugar component in DNA.

DISACCHARIDES AND OLIGOSACCHARIDES

Designation of glycosidic linkages

Identify anomeric hydroxyl group (α or β) and the carbons that are linked. Example: α(1,4) means that carbon 1 in the α position of one monosaccharide is linked to carbon 4 of another monosaccharide.

Digestion

Digestion occurs in the small intestine. If enzymes are deficient, then in the large intestine, they draw in water (causing diarrhea) and/or are fermented by bacteria producing gas (causing bloating and cramps). Example: lactose intolerance caused by reduced lactase synthesis after childhood

Reducing Sugars

If one of the rings can revert to its open-chain form to regenerate the aldehyde, then the sugar is a reducing sugar. Lactose, maltose, and cellobiose are reducing sugars.

Lactose: galactose β(1,4) glucose, found in milk[1]

Maltose: glucose α(1,4) glucose, degradation product of starch, malt sugar

Cellobiose: glucose β(1,4) glucose, degradation product of cellulose

Sucrose: glucose α,β(1,2) fructose, energy source produced in plants. Since the glycosidic bond links both anomeric carbons, sucrose is a nonreducing sugar.

Oligosaccharides

Oligosaccharides are usually attached to polypeptides in glycoproteins and some glycolipids. Example: oligosaccharides attached to membrane and secretory proteins in the endoplasmic reticulum and the Golgi complex of various cells

TWO CLASSES: N-LINKED VS. O-LINKED

N-linked: attached to polypeptides by an N-glycosidic bond with the side chain amide group of Asn; three types of Asn-linked oligosaccharides: high mannose, hybrid, complex

O-linked: attached to polypeptides by the –OH of Ser or Thr, or the –OH of membrane lipids

POLYSACCHARIDES: ENERGY STORAGE OR STRUCTURAL MATERIALS

Hundreds to thousands of sugar units; linear (unbranched) or branched

Homopolysaccharides - made from only one type of monosaccharide

Glycogen and starch: compact structures are ideal for energy storage.

GLYCOGEN: CARBOHYDRATE STORAGE IN VERTEBRATES

Contains more branches than amylopectin: as many as one every 4 at the core of the molecule, and one every 8-12 in the outer regions

Many non-reducing ends gives good access to enzymes, so glucose can be rapidly mobilized from glycogen.

STARCH: CARBOHYDRATE STORAGE IN PLANTS

AMYLOSE forms long, tight left-handed helices, contains several thousand glucose residues, and contains only one reducing end.

AMYLOPECTIN contains a few thousand to 10^6 glucose residues, and its branches prevent helix formation

---

[1] Lactose intolerance is rather normal among adults in the general population. Adults who drink a lot of milk (especially in the US) are the exception.

Iodine test: iodine inserts into helices of amylose and results in blue color

Digestion of starch (initial products are maltose, maltotriose, α-limit dextrins)
1. Mouth: salivary enzyme α-amylase
2. Small intestine: pancreatic α-amylase, other enzymes to convert to glucose
3. Glucose: from small intestine to bloodstream, to liver, then to rest of body

CELLULOSE: THE MOST ABUNDANT ORGANIC SUBSTANCE ON EARTH

Great strength comes from hydrogen bonding between extended linear cellulose chains

Microfibrils = parallel pairs of cellulose molecules (with up to 12,000 glucose residues) held together by hydrogen bonding; bundles of microfibrils contain about 40 pairs

Digestion: only by microorganisms that have cellulase; animals such as termites and cows have these microorganisms in their digestive tracts; if not digested, it's still important as dietary fiber

Summary of Homopolysaccharides

homopolysaccharide (overall shape)		sugar units, glycosidic bonds	function	found in
GLYCOGEN (helical, branched)		α-D-glucose α(1,4) with α(1,6) branches	energy storage	vertebrates (mostly in liver and muscle cells)
STARCH	**AMYLOSE** (left-handed helices)	α-D-glucose α(1,4), unbranched	energy storage	plants
	AMYLOPECTIN (branched every 20-25 residues)	α-D-glucose α(1,4) with α(1,6) branches		
CELLULOSE (extended linear chains with hydrogen bonding between chains)		β-D-glucose β(1,4), unbranched	structural, forms microfibrils and bundles	plants (both primary and secondary cell walls)
CHITIN		N-acetyl glucosamine β(1,4), unbranched	structural	arthropod exoskeletons, fungi cell walls

Chitin: N-acetyl glucosamine

Strong hydrogen bonds between chitin chains; Chitin forms microfibrils

Types of chitin structures:

α-chitin: antiparallel bundles, most stable and common

β-chitin: parallel bundles

γ-chitin: mixture of parallel and antiparallel

β- and γ- are more flexible than α-chitin

Heteropolysaccharides - *MADE FROM TWO OR MORE TYPES OF MONOSACCHARIDES*

Glycosaminoglycans (GAGs) - Principal components of proteoglycans

Linear; composed of disaccharide repeat units that contain a hexuronic acid (except keratan sulfate, which contains galactose)

Classified according to sugar residues, glycosidic linkages, presence and location of sulfate groups

Five classes: hyaluronic acid, chondroitin sulfate, dermatan sulfate, heparin and heparan sulfate, keratan sulfate (see table 7.1 on page 222)

Highly negatively-charged chains repel each other and attract large volumes of water.

Murein (Peptidoglycan)

Major structural component of bacterial cell walls, supplies strength and rigidity

Contains N-acetyl glucosamine (NAG), N-acetyl muramic acid (NAM), and several different amino acids

Three basic components of murein (peptidoglycan)

1. backbone: NAG-NAM disaccharide repeat units linked by β(1,4) glycosidic bonds
2. parallel tetrapeptide chains; each chain attached to N-acetyl muramic acid
3. peptide cross-bridges link the tetrapeptide chains of adjacent molecules

Glycoconjugates: Covalent linkages of carbohydrates to proteins and lipids

Glycolipids = Oligosaccharide-containing lipids

Located on the outer surface of plasma membranes (Chapter 11)

Proteoglycans = GAG chains linked to core proteins

N- and O- glycosidic linkages, high carbohydrate content

Located in extracellular matrix (intercellular material) of tissues

Large numbers of GAGs trap large volumes of water and cations

Contribute support and elasticity to tissues

Example: cartilage - strength, flexibility, resilience; part of meshwork (with matrix proteins such as collagen, fibronectin, laminin) that supports multicellular tissues

Genetic diseases associated with proteoglycan metabolism: mucopolysaccharidoses (example: Hurler's syndrome - dermatan sulfate accumulates)

Glycoproteins = Proteins covalently linked to carbohydrate

N-glycosidic linkage to asparagine or O-glycosidic linkage to serine or threonine

Carbohydrate content varies (1% to more than 85% by weight)

ASPARAGINE-LINKED CARBOHYDRATE

N-glycosidic linkage between N-acetylglucosamine (GlcNAc) and asparagine

Core is constructed on a membrane-bound lipid molecule

Types:

High-mannose type ..... GlcNAc and mannose

Complex type .... GlcNAc and mannose, (may contain fucose, galactose, sialic acid)

Hybrid type .... features of both high-mannose and complex types

MUCIN-TYPE CARBOHYDRATE

O-glycosidic linkage, most common is between N-acetylgalactosamine (GalNAc) and the hydroxyl group of Ser or Thr

Carbohydrate components vary in size and structure. (examples: Gal-β(1,3)-GalNAc, disaccharide found in antifreeze glycoprotein of antarctic fish; complex oligosaccharides of blood groups such as the ABO system)

GLYCOPROTEIN FUNCTIONS AND THE ROLE OF THE CARBOHYDRATE COMPONENT

EXAMPLES OF GLYCOPROTEIN FUNCTIONS:

Metal-transport proteins transferrin and ceruloplasmin

Blood-clotting factors

Complement (proteins involved in cell destruction during immune reactions)

Hormones (example: follicle stimulating hormone (FSH) - stimulates development of eggs and sperm)

Enzymes (example: ribonuclease (RNase) - degrades ribonucleic acid)

Integral membrane proteins {examples: Na^+-K^+-ATPase - ion pump in the plasma membrane of animal cells; major histocompatibility antigens (cell surface markers used to cross-match organ donors and recipients)}

STABILIZES PROTEIN MOLECULES

Protects from denaturation (example: bovine RNase A vs. RNase B)

Increases resistance to proteolysis (carbohydrates on the protein's surface may shield the peptide chains from enzymes)

AFFECTS BIOLOGICAL FUNCTION

Saliva: high viscosity of salivary mucins is due to high content of sialic acid residues

Antifreeze glycoproteins in antarctic fish retard the growth of ice crystals by hydrogen bonding with water molecules

COMPLEX RECOGNITION PHENOMENA

CELL-MOLECULE INTERACTIONS

Insulin receptor - binds to insulin and facilitates transport of glucose into cells, in part by recruiting glucose transporters to the plasma membrane

Glucose transporter - transports glucose into cells

CELL-VIRUS INTERACTIONS

gp120 (target cell binding glycoprotein of HIV) attaches to CD4 receptor on surface of several human cell types (removing oligosaccharides from gp120 reduces its binding to CD4 receptor)

CELL-CELL INTERACTIONS

Cell structure glycoproteins - components of glycocalyx (cell coat), important in cellular adhesion; cell adhesion molecules (CAMs)

SPECIAL INTEREST BOX 7.3: BIOLOGICAL INFORMATION & THE SUGAR CODE

Lectins are carbohydrate binding proteins that are not antibodies, have no enzymatic activity, and consist of 2 or 4 subunits. Lectins "translate the sugar code." They possess recognition domains that bind to specific carbohydrate groups via hydrogen bonds, van der Waals forces, and hydrophobic interactions. Selectins are a family of lectins that act as cell adhesion molecules.

Examples of biological processes that involve lectin binding:

INFECTIONS BY MICROORGANISMS

Bacterial lectins attach the bacteria cell to the host cell by binding to oligosaccharides on the cell's surface (example: gastritis and stomach ulcers)

MECHANISMS OF MANY TOXINS

Lectin-ligand binding initiates endocytosis into the host cell (example: cholera)

PHYSIOLOGICAL PROCESSES SUCH AS LEUKOCYTE ROLLING

THE ORGANIC CHEMISTRY CONNECTION

In organic, you learned that the most stable conformation of a six-membered ring is the chair conformation, and the equatorial positions are more stable than the axial positions. It should come as no surprise, then, that the monosaccharide with all of its substituents in the equatorial positions is β-D-glucose, the most common monosaccharide for both energy storage and as a building block for structural materials, namely, cellulose.

β-D-glucose	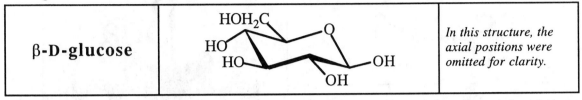	*In this structure, the axial positions were omitted for clarity.*

Remember that the axial positions were always "up" or "down" and the equatorial positions were labeled relative to the axial positions. In D–hexoses, carbon 6 is always drawn "up" in the Haworth projection. So, to draw the Haworth projection of β-D-glucose, first draw the 6-membered ring skeleton with the oxygen in the proper position. Then, add carbon 6 with its OH. The rest of the OH's can then be drawn in following the up-down-up-down-up pattern of the substituents in their equatorial positions. (See the structure in the next section.)

Fischer projections were always shown as a cross, with the carbon understood to be at the center. When the carbon atom was labeled in the center, it was understood that that was not a Fischer projection but a line-bond structure (with no stereochemistry indicated). In this text, however (as in many biochemistry texts), it's understood that any straight-chain carbohydrate drawn vertically (with the most oxidized carbon nearest the top), even with the carbons labeled, is a Fischer projection.

HINTS FOR DRAWING HAWORTH STRUCTURES

Haworth structures show the cyclic (pyranose or furanose) forms of a sugar. Note that for the sake of simplicity, biochemists typically don't include H labels in Haworth projections. This convention lets us focus on the location of the OH groups with less clutter.

- Number the carbons in both the Fischer projection and the Haworth structure.

- First, draw in the atoms attached to the anomeric carbon (that's the first carbon in aldoses and the second carbon in ketoses). For D-sugars, α is "down" and β is "up."*

- The last carbon is always "up" in D-sugars. Draw that one next.

- Imagine the Fischer projection "falling over" to the right. The OH groups that were on the left in the Fischer projection will be "up" in the Haworth structure. Remember that the OH on the carbon farthest from the C=O reacted to form the ring, so that carbon will not have an OH attached to it. In other words, the oxygen within the ring had been an OH, and the anomeric OH group had been the carbonyl oxygen.

- Double-check your work by comparing the numbered carbons on both structures.

- Practice. Using the Fischer projections in Figure 7.3 on p. 203, practice drawing Haworth projections of the aldohexoses. The correct Haworth projections for all of the aldohexoses are at the end of this chapter.

Anomers of D-sugars* : α vs. β

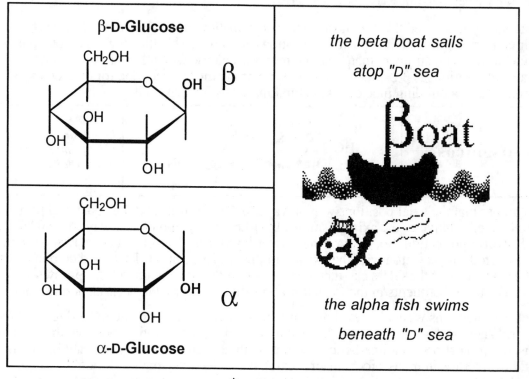

*L-sugars are opposite: α is "up" and β is "down"

HOW MANY STEREOISOMERS ARE POSSIBLE?

Because carbohydrates have many chiral centers, there are many stereoisomers. Identifying which carbons are chiral is the first step in determining the number of stereoisomers. Remember that for a carbon to be chiral, it must have *four different groups* attached to it.

The total number of possible stereoisomers is related to the number of chiral carbons:

number of stereoisomers = 2^n, where n = number of chiral carbons

Why is this true? *Each* chiral carbon can have two different arrangements of groups (that is, they can be mirror images of each other).[2] Glucose is an example of an aldohexose, which has four chiral carbons.

Let's look more closely at glucose (an aldohexose):

1 H—C=O	The first carbon has only three substituents attached to it, and so it's *not* chiral.
2 H—C—OH	The second carbon has four different groups attached to it, so it *is* chiral. The third, fourth, and fifth carbons are also chiral.
3 HO—C—H	
4 H—C—OH	
5 H—C—OH	The sixth carbon has two hydrogens attached to it, and so it's *not* chiral.
6 CH2OH	Glucose, therefore, has four chiral carbons.

---

[2] Organic chemists would differentiate between the two possible arrangements of groups by labeling each chiral carbon atom as *R* or *S*. (Remember them?)

The number of isomers (including glucose) = 2^4 = 2 x 2 x 2 x 2 = 16 possible isomers.

Half of these isomers are enantiomers, or mirror images, and are classified as D- or L-. So, of the 16 possible aldohexose isomers, eight are D-isomers and eight are L-isomers. The structures of all eight D-isomers are shown in Figure 7.3 of your text.

Enantiomers

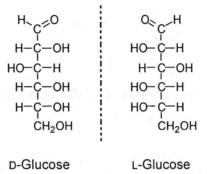

D-Glucose   L-Glucose

The remaining isomers are diastereomers. Diastereomers are stereoisomers, but they are not mirror images of each other.

Let's look at fructose (a ketose) as another example:

CH₂OH C=O HO—C—H H—C—OH H—C—OH CH₂OH	The first carbon has two hydrogens attached to it and is therefore *not* chiral.
	The second carbon has only three groups attached to it and is therefore *not* chiral.
	The third, fourth, and fifth carbons *are* chiral.
	The sixth carbon has two hydrogens attached to it and is therefore *not* chiral.
	Thus, fructose has three chiral carbons, so it has 2^3 = 2 x 2 x 2 = 8 possible isomers (4 D- and 4 L-isomers)

AFTER STUDYING THIS CHAPTER, YOU SHOULD BE ABLE TO SOLVE THESE TYPES OF PROBLEMS:

(R = Review Questions, T = Thought Questions)

- Convert a Fischer projection to a Haworth structure, or draw a Haworth structure given only the name of a compound. (R1, R6, T5)

- Identify pairs of compounds as epimers, anomers, enantiomers, diastereomers, or an aldose-ketose pair. Identify an anomer as α or β. (R2, R11b, R20)

- Distinguish between reducing and nonreducing sugars. (R7, R11c, R16)

- Determine the number of possible stereoisomers of a given sugar. (R10, T6)

- Name a sugar given its structure. Include α or β and appropriate designations for any glycosidic linkages present. (R11a)

- Draw the structure of a di-, tri-, or polysaccharide, given the names of the component mono-(or di-)saccharides and the type(s) of glycosidic linkages present. (T1, T5)

- How do small differences in carbohydrate structure impact its function? (For example, compare cellulose and glycogen.)

CHAPTER 7: ANSWERS TO EVEN-NUMBERED REVIEW QUESTIONS

2. a. Glucose and mannose are examples of epimers.

   b.
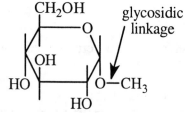

   c. Glucose is a reducing sugar.

   d. Ribose is a monosaccharide.

   e. α-and β-Glucose are anomers.

   f. Glucose and galactose are diastereomers.

4. a. Ribonuclease B is an example of a glycoprotein.

   b. Each proteoglycan contains glycosaminoglycans such as chondroitin sulfate and dermatan sulfate which are linked to a core protein via glycosidic linkages.

   c. Lactose is an example of a disaccharide.

   d. Heparin is a glycosaminoglycan.

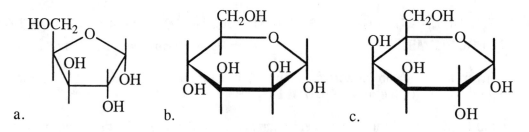

6. a.   b.   c.

8. Starch and glycogen are both homopolysaccharides containing glucose monomers linked by α (1,4) glycosidic bonds with branch points connected by α (1,6) glycosidic bonds. Glycogen, however, is much more highly branched than starch. Cellulose is a linear polymer of glucose linked by β (1,4) glycosidic bonds.

10. Ribulose has 2^2 or 4 possible stereoisomers while sedoheptulose has 2^4 or 16 isomers.

12. a. Glycogen stores glucose.

    b. Glycosaminoglycans are components of proteoglycans.

    c. Glycoconjugates may serve as membrane receptors.

    d. Proteoglycans provide strength, support, and elasticity to tissue.

    e. Hormones such as FSH and enzymes such as RNAse are glycoproteins.

    f. Polysaccharides play important roles in the storage of carbohydrate (starch and glycogen) and the structure of plants (cellulose).

14. In glycoproteins carbohydrate moieties are most frequently linked to asparagine, serine and threonine.

16. A reducing sugar reduces Cu(II) in Benedict's reagent. This reduction takes place because the hemiacetal portion of a sugar can form an aldehyde functional group, which can be oxidized to a carboxylic acid.

18. Numerous carbohydrate groups protect glycoproteins from denaturation.

20. a. D-erythrose and D-threose are epimers.

    b. D-glucose and D-mannose are epimers.

    c. D-ribose and L-ribose are enantiomers.

    d. D-allose and D-galactose are diastereomers.

    e. D-glyceraldehyde and dihydroxyacetone are an aldose-ketose pair.

CHAPTER 7: ANSWERS TO EVEN-NUMBERED THOUGHT QUESTIONS

2. When steroids are conjugated with a uronic acid, the OH groups of the uronic acid form hydrogen bonds with the water. This structural feature increases the solubility of the conjugated steroid molecule.

4. Pathogenic organisms bind to the milk oligosaccharides instead of the oligosaccharides on the surface of the infant's intestinal cells, thus preventing infections.

6. The maximum number of isomers for mannuronic acid is 2^4 or 16 (since mannuric acid has four chiral centers).

8. The water that is absorbed in large quantities by proteoglycans is incompressible. Therefore, proteoglycans provide tissues that contain them in large amounts some protection against mechanical stress, i.e., the tissue resists deformation when pressure is applied.

HAWORTH PROJECTIONS FOR α-D-ALDOHEXOSES *(PYRANOSE FORMS)*

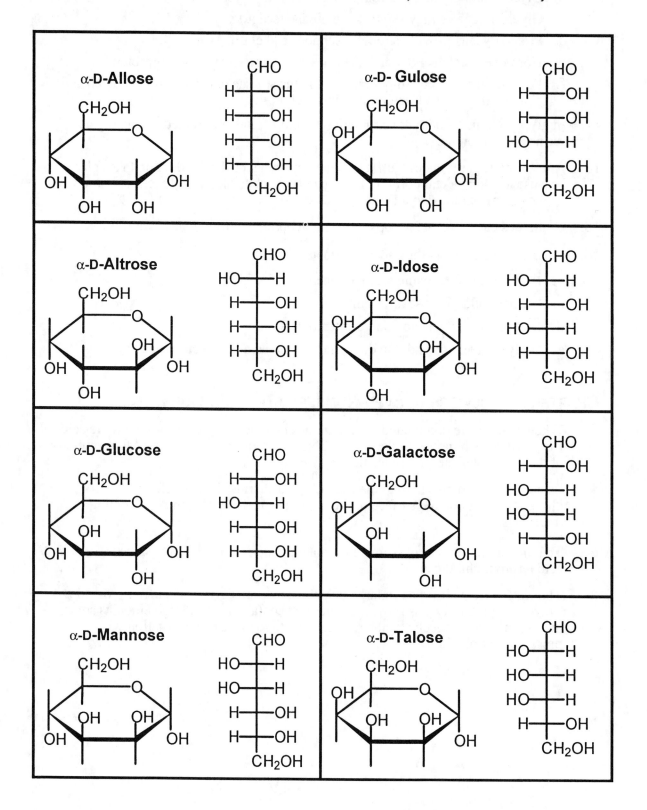

Carbohydrate Metabolism

Studying metabolism brings together what you've learned about structures of biomolecules and how energy flows in living systems.

An important tip in studying metabolism is to key in on the ultimate goal of the pathway and how each reaction brings you closer to the desired end product.

Begin by visualizing the structure of the starting material and follow the changes that occur until you get to the final product. This simplifies remembering the sequence or order of reactions. Focus on the structure of the intermediates. At each step, note any changes that occur to the molecule to understand the kind of reaction that's occurring. And, remember that each enzyme's name clues the type of reaction it catalyzes. For example, a kinase transfers phosphate groups.

Know yourself and your learning styles, and tap into your strengths. If you learn best by reading, create lists to study from. Visual learners will do well to create diagrams of the pathways. Those who learn best by hearing will want to find a study partner to talk through the pathways. Kinesthetic learners learn by doing, and practicing writing out the pathways might be the best bet. Most of us are a combination of the four. Try each style of learning - you might be surprised by which method clicks the best. Studying metabolism lends itself to all four learning styles, so it's a great subject to dive into. (It also gives meaning to everything we've learned so far!)[1]

Example: Glycolysis cleaves glucose to form two 3-carbon molecules. In this pathway, a kinase adds a *second* phosphate group to fructose-6-phosphate. *Why?* After cleavage, both halves will have a phosphate group that traps each half in the cell. Later, those phosphates will be transferred to ADP to form ATP.

Pathway	*General Function*	*Each pathway is activated when:*
Glycolysis	glucose → 2 pyruvate makes 2 ATP, 2 NADH	…energy is needed: either anaerobic (pyruvate forms lactate) or aerobic (pyruvate enters the citric acid cycle)
Gluconeogenesis	pyruvate → glucose requires 4 ATP, 2 GTP	…the liver needs to raise blood sugar levels and liver glycogen is depleted
Glycogenolysis	glycogen → glucose	…muscles need glucose for energy or the liver needs to raise blood sugar levels
Glycogenesis	glucose → glycogen requires 1 UTP/glucose	…glucose is in excess
Pentose Phosphate Pathway	glucose-6-P→ribose-5-P, other sugars makes 2 NADPH	…NADPH is needed for lipid synthesis …ribose is needed for nucleotide synthesis

[1] Visual, Aural, Read/write, and Kinesthetic modes of learning are described on the VARK web page, http://vark-learn.com. VARK is a questionnaire that can be used to discover your learning preferences. If this link is outdated, try a general search on VARK at www.google.com. Copyright for VARK is held by Neil D. Fleming, Christchurch, New Zealand and Charles C. Bonwell, Green Mountain, Colorado, USA.

GLYCOLYSIS: THE OLDEST AND SIMPLEST WAY TO PRODUCE ENERGY

- Glycolysis provides energy (two ATPs), intermediates (two pyruvates), and reducing power (two NADHs).

- It splits one glucose (with 6 carbons) into two pyruvates (with three carbons each).

- It doesn't need O_2 (it's anaerobic).

- It's amphibolic - it functions in both anabolic and catabolic processes.

- It's activated when the cell needs energy, and inhibited when there's plenty of ATP.

The Overall Reaction of Glycolysis: $\Delta G = -103.8$ kJ/mol

D-glucose + 2 ADP + 2 P_i + 2 NAD$^+$ →

2 pyruvate + 2 ATP + 2 NADH + 2H$^+$ + 2 H_2O

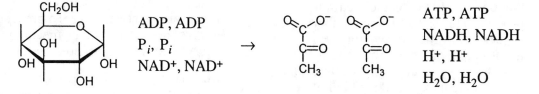

ADP, ADP
P_i, P_i
NAD$^+$, NAD$^+$

ATP, ATP
NADH, NADH
H$^+$, H$^+$
H_2O, H_2O

The Fates of Pyruvate

IF O_2 IS PRESENT: pyruvate → acetyl-CoA → citric acid cycle (energy, CO_2 + H_2O)

The NADH formed may enter the electron transport chain, and ultimately donates its electrons to oxygen, generating more ATP in the process.

IF O_2 IS ABSENT:

The cell needs the ATP from glycolysis for energy. Glycolysis needs a steady supply of NAD$^+$, which must be regenerated from NADH. Otherwise, the cell's NAD$^+$ supply will be depleted and glycolysis would stop.⊗

IN MUSCLE CELLS AND SOME BACTERIA: HOMOLACTIC FERMENTATION

IN YEAST CELLS: ALCOHOLIC FERMENTATION

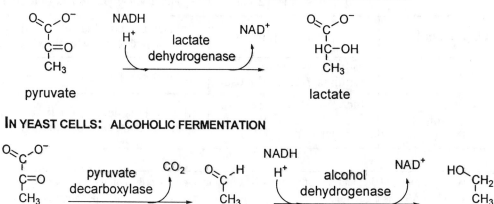

GLYCOLYSIS

STAGE ONE: GLUCOSE IS PHOSPHORYLATED AND CLEAVED TO FORM TWO G-3-P

Glycolysis occurs only in the cytoplasm.
The three irreversible reactions - hexokinase, PFK-1, and pyruvate kinase (in stage two) - are highly regulated.

1. Hexokinase, Mg^{2+}

Charged molecules can't pass
   through the cell membrane
Investment of one ATP
IRREVERSIBLE!
   Inhibited by ATP and
   glucose-6-phosphate (except for
   liver hexokinase D)

2. Phosphoglucose isomerase

Now C-1 can be phosphorylated
Enediol intermediate

3. Phosphofructokinase-1, Mg^{2+}

Now both halves will have a
   phosphate after cleavage
Phosphates will make ATP later
IRREVERSIBLE!
   Activated by fructose-2,6-
   bisphosphate
   Inhibited by ATP and citrate

4. Aldolase

Cleavage
Reverse aldol condensation

5. Triose phosphate isomerase

Now the DHAP half won't be wasted

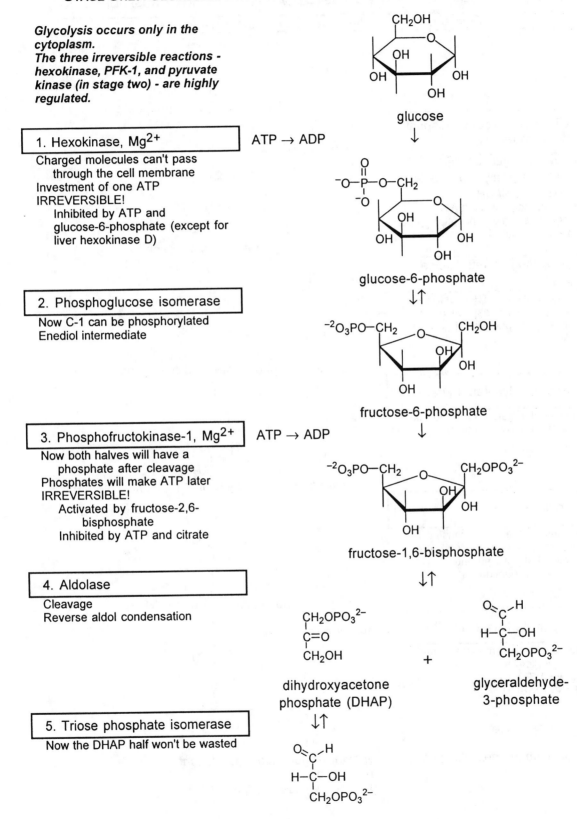

glucose

$ATP \rightarrow ADP$

glucose-6-phosphate

fructose-6-phosphate

$ATP \rightarrow ADP$

fructose-1,6-bisphosphate

dihydroxyacetone phosphate (DHAP) + glyceraldehyde-3-phosphate

GLYCOLYSIS

STAGE TWO: ENERGY RECAPTURE STAGE (OXIDATION-REDUCTION-PHOSPHORYLATION)

*From this point on, there are **two** of each molecule below.*

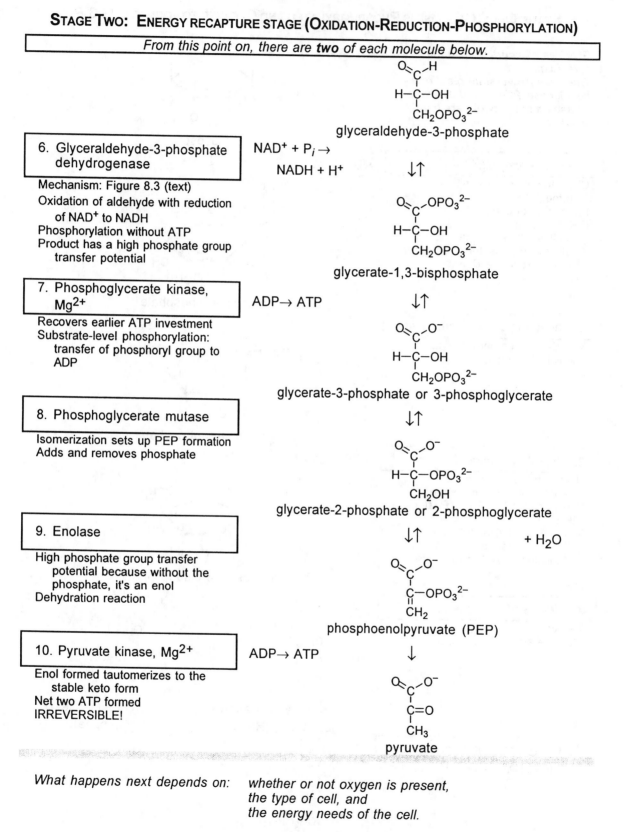

6. Glyceraldehyde-3-phosphate dehydrogenase

Mechanism: Figure 8.3 (text)
Oxidation of aldehyde with reduction
  of NAD^+ to NADH
Phosphorylation without ATP
Product has a high phosphate group
  transfer potential

glyceraldehyde-3-phosphate

$NAD^+ + P_i \rightarrow$
$\quad$ NADH + H$^+$

glycerate-1,3-bisphosphate

7. Phosphoglycerate kinase, Mg^{2+}

Recovers earlier ATP investment
Substrate-level phosphorylation:
  transfer of phosphoryl group to
  ADP

$ADP \rightarrow ATP$

glycerate-3-phosphate or 3-phosphoglycerate

8. Phosphoglycerate mutase

Isomerization sets up PEP formation
Adds and removes phosphate

glycerate-2-phosphate or 2-phosphoglycerate

9. Enolase

High phosphate group transfer
  potential because without the
  phosphate, it's an enol
Dehydration reaction

$+ H_2O$

phosphoenolpyruvate (PEP)

10. Pyruvate kinase, Mg^{2+}

Enol formed tautomerizes to the
  stable keto form
Net two ATP formed
IRREVERSIBLE!

$ADP \rightarrow ATP$

pyruvate

What happens next depends on: whether or not oxygen is present,
the type of cell, and
the energy needs of the cell.

GLUCONEOGENESIS: THE GENESIS OF NEW GLUCOSE

- is an *anabolic* pathway used to synthesize glucose from amino acids or lactate.

- requires energy: the hydrolysis of 4 ATP and 2 GTP.

- is the reverse of glycolysis except for reactions that bypass the three irreversible reactions of glycolysis

- is important because the brain and red blood cells need a steady supply of glucose (use glucose as their primary energy source).

- occurs primarily in the liver after glycogen is depleted

The overall reaction of gluconeogenesis: pyruvic acid to glucose

$$2\ C_3H_4O_3 + 4\ ATP + 2\ GTP + 2\ NADH + 2\ H^+ + 6\ H_2O \rightarrow$$
$$1\ C_6H_{12}O_6 + 4\ ADP + 2\ GDP + 2\ NAD^+ + 6\ HPO_4^{2-} + 6\ H^+$$

Gluconeogenesis Substrates

LACTATE: CORI CYCLE:

Exercising muscle:.... glucose to pyruvate to lactate (glycolysis/lactic acid fermentation); lactate enters bloodstream

Transport to liver:..... lactate to pyruvate to glucose (gluconeogenesis)

GLYCEROL → GLYCEROL-3-PHOSPHATE → DIHYDROXYACETONE PHOSPHATE (DHAP)

This occurs only in the liver, when cytoplasmic $[NAD^+]$ is high; uses an ATP but creates an NADH. Glycerol is a product of fat catabolism.

GLUCOGENIC AMINO ACIDS

ALANINE CYCLE:

In exercising muscle, alanine transaminase transfers $-NH_3^+$ from glutamate to pyruvate:

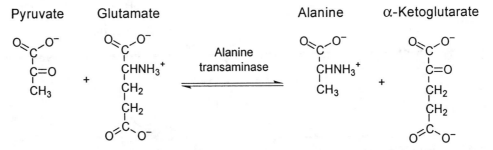

Alanine is transported to the liver, where it's converted back to pyruvate again via alanine transaminase in the reverse reaction.

REACTIONS OF GLUCONEOGENESIS THAT ARE NOT THE REVERSE OF GLYCOLYSIS *(AND SO USE DIFFERENT ENZYMES)*

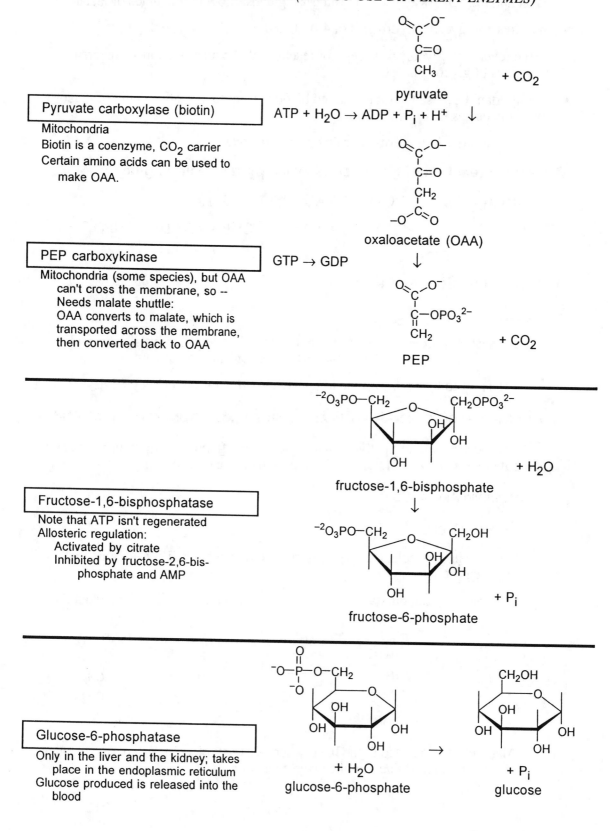

Pyruvate carboxylase (biotin)

Mitochondria
Biotin is a coenzyme, CO_2 carrier
Certain amino acids can be used to make OAA.

pyruvate

$ATP + H_2O \rightarrow ADP + P_i + H^+$ $\downarrow$

oxaloacetate (OAA)

PEP carboxykinase

Mitochondria (some species), but OAA can't cross the membrane, so --
Needs malate shuttle:
OAA converts to malate, which is transported across the membrane, then converted back to OAA

$GTP \rightarrow GDP$

$\downarrow$

PEP $+ CO_2$

$+ CO_2$

fructose-1,6-bisphosphate

$+ H_2O$

$\downarrow$

Fructose-1,6-bisphosphatase

Note that ATP isn't regenerated
Allosteric regulation:
   Activated by citrate
   Inhibited by fructose-2,6-bis-
      phosphate and AMP

fructose-6-phosphate

$+ P_i$

Glucose-6-phosphatase

Only in the liver and the kidney; takes place in the endoplasmic reticulum
Glucose produced is released into the blood

glucose-6-phosphate

$+ H_2O$

$\rightarrow$

glucose

$+ P_i$

Summary of Regulation of Glycolysis and Gluconeogenesis

COMPARTMENTATION:

Glycolysis occurs only in the cytoplasm, and several gluconeogenesis reactions occur in the mitochondria. Glycerol kinase is found only in the liver.

REGULATION BASED UPON THE NEEDS OF THE CELL:

Keep in mind the ultimate goal of each pathway, and their regulation will make sense. For example, glycolysis produces energy, reducing power, and intermediates. So, when the cell needs energy (when reserves are low), glycolysis is activated and gluconeogenesis is inhibited. When the cell has enough energy, glycolysis is inhibited and gluconeogenesis is activated. *Substrate cycle:* paired reactions that are coordinately regulated

THE LIVER MODULATES BLOOD SUGAR LEVELS

When blood sugar is low, the liver releases glucose into the bloodstream, first by degrading glycogen, then by gluconeogenesis (in response to glucagon).

When blood sugar is high, insulin inhibits gluconeogenesis and activates glycolysis in the liver. Also, hexokinase D in the liver isn't inhibited by glucose-6-phosphate, so it lets the liver remove glucose from the blood to store as glycogen.

Regulation by Covalent Modification

PFK-2 and fructose-2,6-bisphosphatase is actually one enzyme that's bifunctional: without its phosphate, it's PFK-2, and with its phosphate, it's fructose-2,6-bisphosphatase.

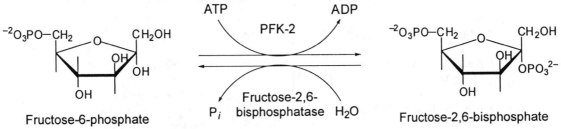

Fructose-6-phosphate   Fructose-2,6-bisphosphate

High Blood Sugar:

Insulin activates PFK-2, which increases fructose-2,6-bisphosphate levels, which:

- activates PFK-1 (glycolysis) in the liver
- inhibits fructose-2,6-bisphosphatase (inhibits gluconeogenesis)

Regulation of Glycolysis and Gluconeogenesis

When Energy is Needed	To Store Energy or To Raise Blood Glucose
ACTIVATES GLYCOLYSIS	ACTIVATES GLUCONEOGENESIS
Glucose-6-Phosphate	2 Pyruvate (also: lactate, glycerol, some amino acids)
⇩	⇩
2 Pyruvate	Glucose-6-Phosphate

PFK-1 activated by:
   AMP
   Fructose-2,6-bisphosphate
 	
Pyruvate kinase activated by:
   AMP
   Fructose-1,6-bisphosphate
   (feed-forward control)
 	
Insulin stimulates the synthesis of:
   gluckokinase, PFK-1, PFK-2
	Pyruvate carboxylase activated by:
	   Acetyl CoA (high during starvation;
	product of fatty acid catabolism)
	
	Fructose-1,6-bisphosphatase
	   activated by:
	   ATP
	
	Glucagon stimulates the synthesis of
	   PEP carboxykinase,
	   fructose-1,6-bisphosphatse,
	   glucose-6-phosphatase.

INHIBITS GLUCONEOGENESIS
 	
Pyruvate carboxylase inhibited by:
   Acetyl-CoA
 	
Fructose-1,6-bisphosphatase
   inhibited by:
   AMP
   Fructose-2,6-bisphosphate
 	
(Fructose-2,6-bisphosphate
indicates high levels of glucose.)
	INHIBITS GLYCOLYSIS
	
	Hexokinase inhibited by:
	   ATP
	   Glucose-6-phosphate
	
	PFK-1 inhibited by:
	   ATP
	   Citrate
	
	Pyruvate kinase inhibited by:
	   ATP
	   Acetyl-CoA

THE PENTOSE PHOSPHATE PATHWAY

ULTIMATE GOALS:

1. to make NADPH (reducing power) for syntheses and to help prevent oxidative damage (it's a great reducing agent)

2. to make sugar intermediates, especially ribose-5-phosphate, a component of nucleotides and nucleic acids

LOCATION:

Cytoplasm; especially in cells which synthesize lipids and cells that are at high risk for oxidative damage

In plants, during the dark reactions of photosynthesis

OXIDATIVE PHASE: **ALL THREE REACTIONS ARE IRREVERSIBLE.**

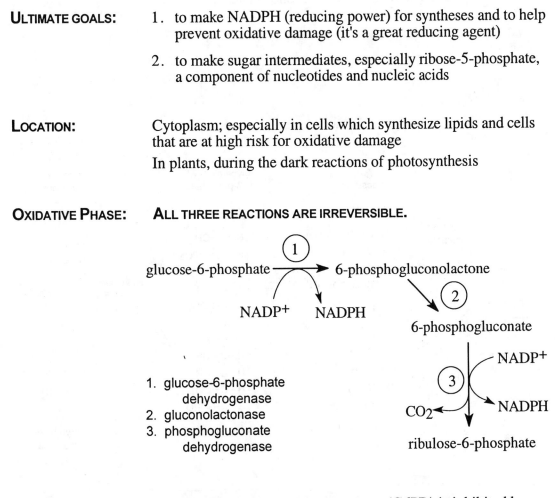

1. glucose-6-phosphate dehydrogenase
2. gluconolactonase
3. phosphogluconate dehydrogenase

REGULATION:

Glucose-6-phosphate dehydrogenase (G6PD) is inhibited by NADPH and activated by glucose-6-phosphate and GSSG, which indicates oxidative damage and a need for NADPH.

A high carbo diet triggers the synthesis of the enzymes G6PD and phosphogluconate dehydrogenase.

NONOXIDATIVE PHASE: **ALL REACTIONS ARE REVERSIBLE.**

What happens to ribulose-5-phosphate depends on the metabolic needs of the cells.

1. If the cell needs ribose-5-phosphate for nucleotide biosynthesis, ribose-5-phosphate isomerase will isomerize ribulose-5-phosphate.

2. If the cell only needs reducing power (NADPH), then the carbons of ribulose-5-phosphate will be recycled and converted into fructose-6-phosphate and glyceraldehyde-3-phosphate, intermediates of glycolysis. To completely recycle the ribulose-5-phosphate carbons via glycolysis, at least three molecules of ribulose-5-phosphate are needed initially. See the reactions outlined below.

3. If the cell needs ribose-5-phosphate but doesn't need reducing power (NADPH), the cell can also use the reverse reactions from the nonoxidative phase, starting with two fructose-6-phosphates and one glyceraldehyde-3-phosphate to make 3 ribose-5-phosphates.

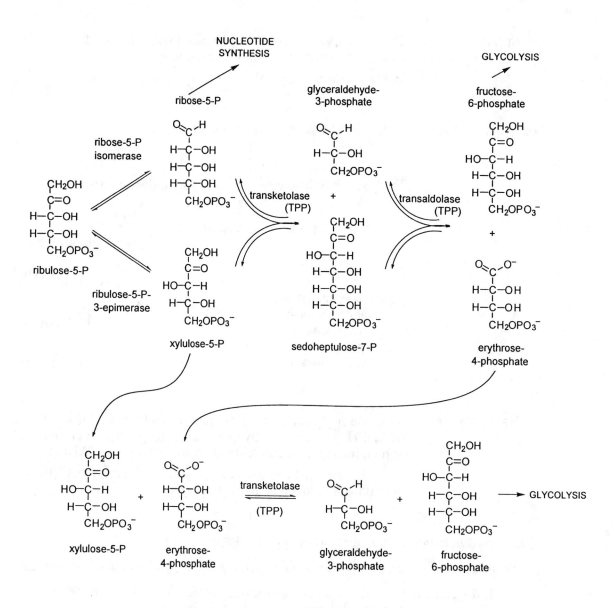

TRANSKETOLASE: TRANSFERS 2 CARBONS FROM A KETO SUGAR TO AN ALDO SUGAR

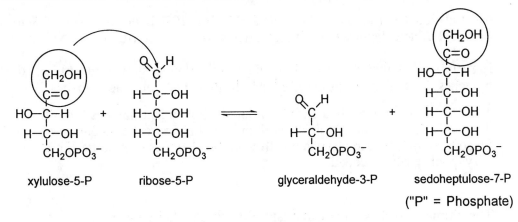

xylulose-5-P   ribose-5-P   glyceraldehyde-3-P   sedoheptulose-7-P

("P" = Phosphate)

TRANSALDOLASE: TRANSFERS 3 CARBONS FROM A KETO SUGAR TO AN ALDO SUGAR

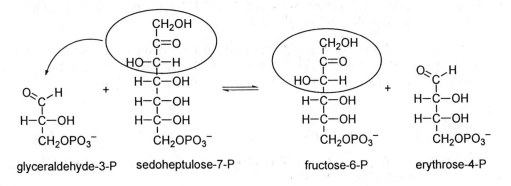

glyceraldehyde-3-P   sedoheptulose-7-P   fructose-6-P   erythrose-4-P

METABOLISM OF OTHER IMPORTANT SUGARS

Fructose Metabolism

LIVER: Four enzymes convert fructose to two G-3-P molecules. Fructose is metabolized quicker than glucose because this bypasses the regulatory enzymes hexokinase and PFK-1.

MUSCLE, ADIPOSE: Hexokinase phosphorylates fructose (to fructose-6-phosphate).

Galactose Metabolism Several reactions ultimately form UDP-glucose for glycogenesis (as is) or glycolysis (after conversion to glucose-6-phosphate).

Mannose Metabolism Hexokinase phosphorylates mannose (to mannose-6-phosphate) then phosphomannose isomerase converts it to fructose-6-phosphate.

GLYCOGEN METABOLISM

Most glycogen is found in the muscle and liver ($\approx$10% liver mass, and $\approx$1% muscle mass). The actual amount of glycogen depends on the nutritional state of the organism.

Remember that glycogen is made of α(1,4)-linked glucose with α(1,6)-linked branches. Special enzymes are needed to form and degrade the branches.

The ends of all of those branches are nonreducing ends. Enzymes for both glycogenesis and glycogenolysis work only on the nonreducing ends. One huge glycogen molecule can be coated with working enzymes so glucose can be released very quickly when needed.

The ultimate goals of GLYCOGENESIS depends on where it takes place:

MUSCLE: Excess glucose is stored as glycogen: ENERGY STORAGE.

LIVER: Glucose is removed from the blood in response to insulin, which signals high blood glucose levels.

Glycogen synthase, and thus glycogenesis, is activated by glucose-6-phosphate, an indicator of excess glucose. High levels of ATP (as well as glucose-6-phosphate) inhibit the glycogenolysis by inhibiting glycogen phosphorylase.

The ultimate goals of GLYCOGENOLYSIS also depends on where it takes place:

MUSCLE: Glycogen is degraded to produce glucose-6-phosphate for ENERGY (glycolysis and beyond).

LIVER: Glycogen is degraded to produce free glucose for EXPORT (to other cells), so glucose-6-phosphate is dephosphorylated and sent to the blood. Glucagon signals low blood sugar, and epinephrine signals an immediate need for energy.

Glycogen phosphorylase, and thus glycogenolysis, is activated by AMP, an indicator that energy is needed.

As with glycolysis and gluconeogenesis, glycogen degradation is not simply a reverse of glycogenesis.

The next two pages outline the reactions, and further notes on the regulation of these two pathways follow.

GLYCOGENESIS: GENESIS OF GLYCOGEN
BEGINNING WITH GLUCOSE-6-PHOSPHATE

Other pathways to glycogen exist, namely glucose to C_3 molecules to liver glycogen.

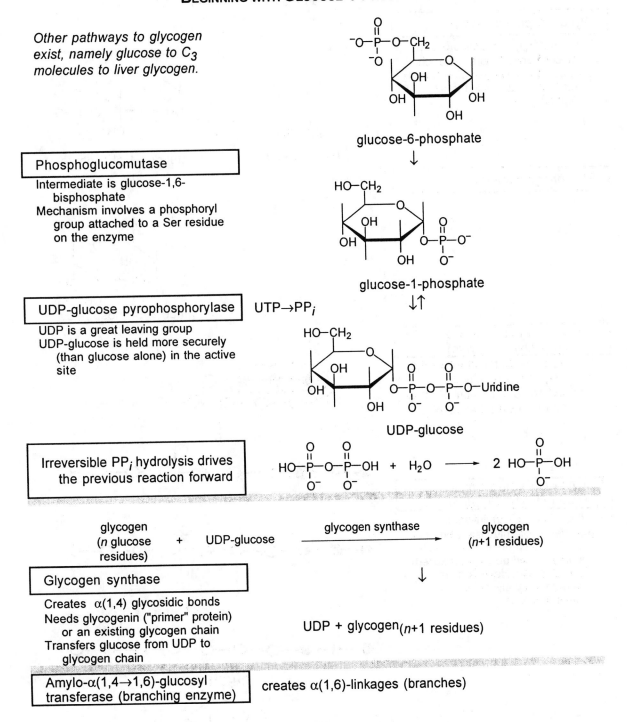

glucose-6-phosphate

Phosphoglucomutase

Intermediate is glucose-1,6-bisphosphate

Mechanism involves a phosphoryl group attached to a Ser residue on the enzyme

glucose-1-phosphate

UDP-glucose pyrophosphorylase   UTP→PP$_i$

UDP is a great leaving group

UDP-glucose is held more securely (than glucose alone) in the active site

UDP-glucose

Irreversible PP$_i$ hydrolysis drives the previous reaction forward

$$glycogen\ (n\ glucose\ residues)\ +\ UDP\text{-}glucose\ \xrightarrow{glycogen\ synthase}\ glycogen\ (n+1\ residues)$$

Glycogen synthase

Creates α(1,4) glycosidic bonds

Needs glycogenin ("primer" protein) or an existing glycogen chain

Transfers glucose from UDP to glycogen chain

UDP + glycogen$(n+1$ residues)

Amylo-α(1,4→1,6)-glucosyl transferase (branching enzyme)   creates α(1,6)-linkages (branches)

GLYCOGENOLYSIS

$$\text{glycogen (} n \text{ residues)} + P_i \; (HPO_4{}^{2-})$$

$$\downarrow$$

Glycogen phosphorylase

Cleaves $\alpha(1,4)$ linkages; stops 4 glucose units before a branch
Limit dextrin - glycogen molecule that's been degraded to its branch points

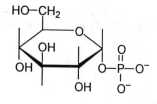

glycogen   +   glucose-1-phosphate

(*n*-1 residues)

Phosphoglucomutase

$$\downarrow\uparrow$$

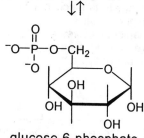

glucose-6-phosphate

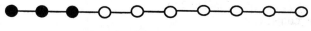

Amylo-α(1,6)-glucosidase (debranching enzyme)

Transfers the outer three of the four glucose residues attached to a branch point to a nearby nonreducing end.

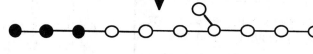

Amylo-α(1,6)-glucosidase (debranching enzyme)

Hydrolysis of α(1,6)-linkages
Removes single glucose residue at each branch point
The unbranched polymer produced by 1,6-glucosidase is then degraded by glycogen phosphorylase.

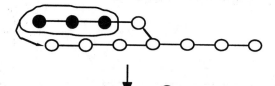

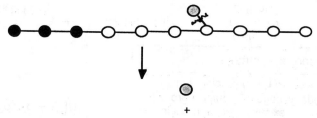

+

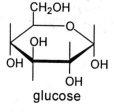

glucose

Regulation of Glycogen Metabolism

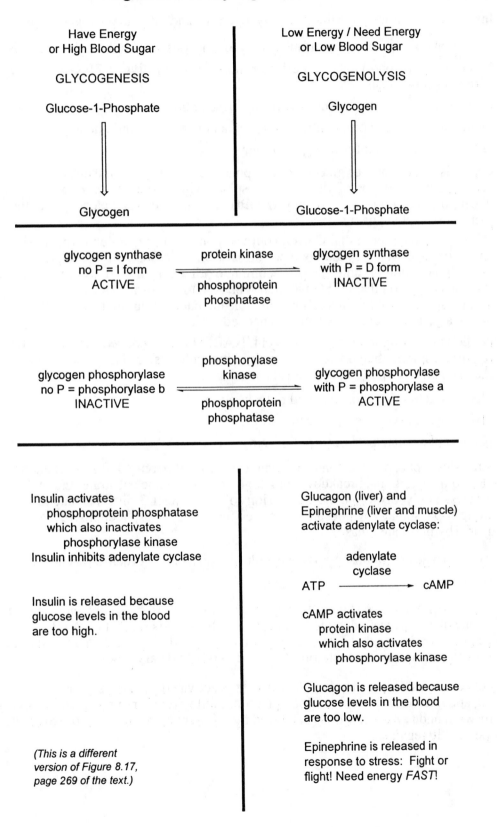

Have Energy or High Blood Sugar	Low Energy / Need Energy or Low Blood Sugar
GLYCOGENESIS	GLYCOGENOLYSIS
Glucose-1-Phosphate	Glycogen
⇓	⇓
Glycogen	Glucose-1-Phosphate

glycogen synthase
no P = I form
ACTIVE

protein kinase

glycogen synthase
with P = D form
INACTIVE

phosphoprotein
phosphatase

glycogen phosphorylase
no P = phosphorylase b
INACTIVE

phosphorylase
kinase

glycogen phosphorylase
with P = phosphorylase a
ACTIVE

phosphoprotein
phosphatase

Insulin activates
  phosphoprotein phosphatase
    which also inactivates
      phosphorylase kinase
Insulin inhibits adenylate cyclase

Insulin is released because
glucose levels in the blood
are too high.

*(This is a different
version of Figure 8.17,
page 269 of the text.)*

Glucagon (liver) and
Epinephrine (liver and muscle)
activate adenylate cyclase:

adenylate
cyclase

ATP ⟶ cAMP

cAMP activates
  protein kinase
    which also activates
      phosphorylase kinase

Glucagon is released because
glucose levels in the blood
are too low.

Epinephrine is released in
response to stress: Fight or
flight! Need energy *FAST*!

CHAPTER 8: ANSWERS TO EVEN-NUMBERED REVIEW QUESTIONS

2. a. Insulin – a hormone that stimulates glycogenesis and inhibits glycogenolysis.

    b. Glucagon – a hormone that stimulates glycogenolysis and inhibits glycogenesis.

    c. Fructose-2,6-bisphosphate – an effector molecule that activates PFK- 1 and stimulates glycolysis.

    d. Congeners – molecules produced during fermentation

    e. Glutathione – an important intracellular reducing agent (or antioxidant)

    f. GSSG – the oxidized form of glutathione

    GSSG is an activator of glucose-6-phosphate dehydrogenase, the enzyme that catalyzes the oxidation of glucose-6-phosphate with $NADP^+$ to produce 6-phospho-D-glucono-γ-lactone and NADPH. This is a key regulatory step in the pentose phosphate pathway.

    (Note: It makes sense that GSSG would activate G-6-PD, which forms NADPH, an antioxidant. When glutathione performs as an antioxidant, it would prevent oxidation of another molecule by becoming oxidized itself, and becoming GSSG. It stands to reason that GSSG would not be able to perform as an antioxidant. The presence of GSSG in a cell would indicate that more antioxidant activity is necessary, and that more NADPH is needed.

    g. NADPH – an important reducing agent. NADPH is the reduced form of NADP+, required for reductive processes (e.g. lipid biosynthesis) and antioxidant mechanisms (NADPH is a powerful antioxidant)

4. a. gluconeogenesis – cytoplasm and mitochondria

    b. glycolysis - only in the cytoplasm

    c. pentose phosphate pathway - cytoplasm.

6. Substrate-level phosphorylation is the synthesis of ATP (from ADP and P_i) that is coupled to the exergonic breakdown of a high energy organic substrate. Examples of this process in glycolysis are the conversions of glycerate-1,3-bisphosphate to glycerate-3-phosphate (by phosphoglycerate kinase) and phosphoenolpyruvate to pyruvate (by pyruvate kinase).

8. Under anaerobic conditions, pyruvate is reduced to lactate in order to regenerate NAD^+.

10. Glycolysis occurs in two stages. In stage I, glucose is phosphorylated and cleaved to two molecules of glyceraldehyde-3-phosphate. During this stage, two ATP molecules are consumed. In stage 2, each glyceraldehyde-3-phosphate is converted to pyruvate, a process in which four ATP and two NADH are produced.

12. Gluconeogenesis occurs mainly in the liver. It is activated by processes such as fasting and exercise that deplete blood glucose. Futile cycles are prevented by having the forward and reverse reactions catalyzed by different enzymes, both of which are independently regulated.

CHAPTER 8: ANSWERS TO EVEN-NUMBERED THOUGHT QUESTIONS

2.   In the synthesis of new glycogen molecules, a primer protein called glycogenin is used to initiate glycogen formation. Glucose is transferred from UDP-glucose to a specific tyrosine residue of the glycogenin. This glucose then serves as the starting point for a new growing glycogen molecule.

4.   Phosphoenolpyruvate has a high phosphate group transfer potential because the transfer of the enol phosphate to another molecule produces a vinyl alcohol. The vinyl alcohol tautomerizes rapidly to the keto-form making the transfer almost irreversible.

6.   If gluconeogenesis and glycolysis were exactly the reverse of one another, futile cycles would be established and much energy would be wasted. In addition, it would be impossible for the body to store glycogen or release glucose into the blood as needed.

CHAPTER NINE

Aerobic Metabolism I: The Citric Acid Cycle

As biomolecules are oxidized to CO_2, they transfer electrons to NAD^+ (and FAD), which is then reduced to NADH and $FADH_2$. In turn, they transfer electrons to the electron transport chain and ultimately to oxygen, which is reduced to H_2O. The energy released during this process pumps H^+ ions to the other side of the membrane. When they naturally flow back in, they are (typically) forced to pass through special channels that synthesize ATP. So, ATP captures energy that results from the transfer of electrons from fuel molecules to (ultimately) oxygen.

OXIDATION-REDUCTION REACTIONS

Review the previous discussions of redox reactions in Chapter 1: page 19 of your text and page 4 of this study guide.

$$\Delta G^{\circ\prime} = - n \, F \, \Delta E_0'$$ (n = number of electrons transferred, F = 96.5 kJ/Vmol)

For redox reactions to be spontaneous (have a negative $\Delta G^{\circ\prime}$), they must have a positive $\Delta E_0'$ (refer to the equation above; n and F will always be positive).

Electrons spontaneously transfer from the more negative E_0' to the more positive E_0'. Think on this: For the half-reaction ($NO_3^- + 2\,H^+ + 2\,e^- \rightarrow NO_2^- + H_2O$), $E_0' = + 0.44$ V. That's why nitric acid tends to oxidize organic compounds - its E_0' is relatively positive, so it's happy to take electrons from other molecules.

When combining two half reactions, make sure that the overall equation is balanced in the number of electrons transferred.[1] When you reverse a half-reaction, also reverse the sign of its E_0'. To calculate $\Delta E_0'$, add the two values of E_0' (including the sign-change of the reaction that you reversed).[2]

Let's look more closely at the example used in Chapter 1. The two half reactions with their E_0' values (as listed in Table 9.1 of your text) are:

$NAD^+ + H^+ + 2\,e^- \rightarrow$ NADH	–0.32 V
Acetaldehyde + $2\,H^+ + 2\,e^- \rightarrow$ ethanol	–0.20 V

For the $\Delta E_0'$ to be positive, and for the electrons to balance on both sides of the overall equation, we must reverse the first reaction:

NADH $\rightarrow NAD^+ + H^+ + 2\,e^-$	+0.32 V
Acetaldehyde + $2\,H^+ + 2\,e^- \rightarrow$ ethanol	–0.20 V
Acetaldehyde + H^+ + NADH $\rightarrow$ ethanol + NAD^+	+0.12 V[3]

---

[1] Remember: In calculating $\Delta E_0'$, don't multiply the value of E_0' by the number of electrons transferred or by the coefficient in the balanced equation. This is different from ΔG° and ΔH° calculations.

[2] This method avoids the confusion of which E_0' to subtract from which. If you feel more comfortable with a different method, by all means, use it.

[3] We've already seen this reaction as the second step of homolactic fermentation, catalyzed by alcohol dehydrogenase. (This is also hiding out in Question 9.2, page 277 of your text.)

To Identify Oxidizing Agents and Reducing Agents:

1. Locate each conjugate redox pair. It often helps to draw the molecular structures.

   In the equation above, the two conjugate redox pairs are acetaldehyde/ethanol and NAD+/NADH.

2. In each conjugate redox pair, one member will be its oxidized form, and the other will be its reduced form. Label them. Remember that the oxidized form will have more oxygens (and/or less H's, and a higher oxidation state), and the reduced form will have less oxygens (and/or more H's, and a lower oxidation state).

NAD+ = oxidized form	*acetaldehyde = oxidized form*
NADH = reduced form	*ethanol = reduced form*

3. Check your labels using the balanced equation. Each side of the equation should have one reduced form and one oxidized form.

$$NADH + H_3C-\overset{\displaystyle O}{\underset{\displaystyle H}{C}} + H^+ \longrightarrow NAD^+ + CH_3CH_2OH$$

reduced form	*oxidized form*	*oxidized form*	*reduced form*

4. Of the two reactants, the reduced form is the reducing agent - it gives away its extra electrons, reducing the other guy (and becoming oxidized in the process). Conversely, the oxidized form is the oxidizing agent - it grabs electrons from the other guy, oxidizing it (and becoming reduced in the process).

CITRIC ACID CYCLE

THE GOALS OF THE CITRIC ACID CYCLE ARE TO:

1. Convert fuel molecules to **ENERGY**: Produce reducing power (3 NADH and 1 $FADH_2$ per acetyl-CoA), which can enter the electron transport chain and provide the energy to synthesize ATP. One GTP (which can easily be converted to ATP) is produced directly, although this is a small energy bonus compared to the number of ATP that are ultimately produced from the electron transport chain.

2. Produce **INTERMEDIATES**: The carbon precursors for lipids, sugars, amino acids, and nucleic acids are all derived from citric acid cycle intermediates.

OVERALL REACTION

Acetyl-CoA + 3 NAD+ + FAD + GDP + P_i + 2 H_2O →

$$2\ CO_2 + 3\ NADH + FADH_2 + CoASH + GTP + 3\ H^+$$

In eukaryotes, the citric acid cycle takes place in the mitochondria. Remember that glycolysis took place in the cytoplasm.

COENZYMES

THIAMINE PYROPHOSPHATE (TPP) - decarboxylates and transfers aldehyde groups; used by pyruvate dehydrogenase (E1 of the pyruvate dehydrogenase complex) and the α-ketoglutarate dehydrogenase complex

LIPOIC ACID - carries hydrogens or acetyl groups; used by dihydrolipoyl transacetylase (E2 of the pyruvate dehydrogenase complex) and dihydrolipoyl transsuccinylase of the α-ketoglutarate dehydrogenase complex

NAD⁺ - carries electrons; used by dihydrolipoyl dehydrogenase (E3 of the pyruvate dehydrogenase complex and of the α-ketoglutarate dehydrogenase complex), isocitrate dehydrogenase, and malate dehydrogenase

FAD - carries electrons; used by dihydrolipoyl dehydrogenase (E3 of the pyruvate dehydrogenase complex and of the α-ketoglutarate dehydrogenase complex), succinate dehydrogenase

COENZYME A (CoASH) - carries acetyl groups via a thioester bond (Acetyl-CoA); used by dihydrolipoyl transacetylase (E2 of the pyruvate dehydrogenase complex) and dihydrolipoyl transsuccinylase of the α-ketoglutarate dehydrogenase complex

ACETYL-CoA

Acetyl-CoA is an acetyl group bound to coenzyme A by a thioester bond. When studying reaction mechanisms that involve acetyl-CoA, it's helpful to remember that the linkage is a thioester. Compare the structure of acetyl-CoA below with the structure of CoASH in Figure 9.6, p. 281 of your text.

{Note that coenzyme A is abbreviated CoASH. To be consistent, instead of "acetyl-CoA" we should probably say "acetyl-S-CoA" (or even "CoAS-acetyl"), but, acetyl-CoA is just easier, and is in common usage.}

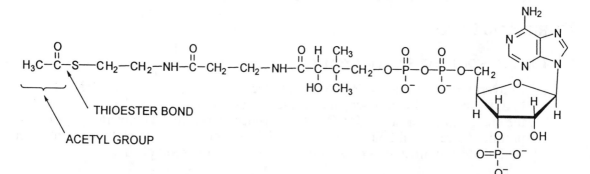

Conversion of Pyruvate to Acetyl-CoA $\Delta G^{\circ\prime} = -33.5$ kJ/mol

OVERALL REACTION: AN OXIDATIVE DECARBOXYLATION

pyruvate + NAD⁺ + CoASH → acetyl-CoA + NADH + CO_2 + H_2O + H⁺

PYRUVATE DEHYDROGENASE COMPLEX = MULTIPLE COPIES OF:

E_1: pyruvate dehydrogenase (or pyruvate decarboxylase); needs TPP

E_2: dihydrolipoyl transacetylase; needs lipoic acid and CoASH

E_3: dihydrolipoyl dehydrogenase; needs FAD and NAD⁺

MECHANISM OF THE PYRUVATE DEHYDROGENASE COMPLEX

(This is simplified to emphasize the changes to the acetyl group. For details regarding the linkage to the TPP thiazole ring, see Figure 9.8 of your text.)

E_1: Decarboxylation of pyruvate to form hydroxyethyl-TPP (HETPP):

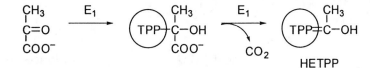

E_2 converts the hydroxyethyl group of HETPP to acetyl-CoA:

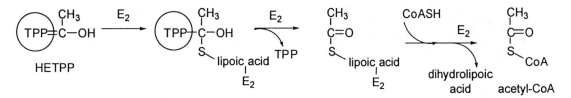

E_3 regenerates lipoic acid; FAD is also regenerated spontaneously.

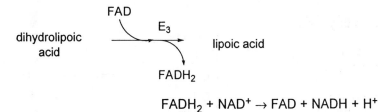

$$FADH_2 + NAD^+ \rightarrow FAD + NADH + H^+$$

REGULATION:

In general, inhibitors are products and indicators that the cell has plenty of energy (e.g., ATP), while activators are substrates and indicators that the cell needs energy (e.g., AMP)

PRODUCT INHIBITION: Acetyl-CoA and NADH are inhibitors.

ALLOSTERIC: ATP inhibits; AMP, NAD^+, CoASH are activators.

COVALENT MODIFICATION: The products acetyl-CoA and NADH activate a kinase, which phosphorylates and inactivates the pyruvate dehydrogenase complex. Substrates pyruvate, CoASH, and NAD^+ inhibit this kinase, removing an inhibitor (that's not quite the same as activating a reaction, but it helps). Low levels of ATP activate a phosphoprotein phosphatase, which dephosphorylates and activates the pyruvate dehydrogenase complex.

Reactions of the Citric Acid Cycle take place in the mitochondrial matrix

Acetyl-CoA brings in two carbons, and two carbons leave as CO_2.

Oxaloacetate is regenerated so the cycle can continue.

THE CITRIC ACID CYCLE

ACETYL-CoA BRINGS IN TWO CARBONS, TWO CARBONS LEAVE AS CO_2

Acetyl-CoA enters the cycle.

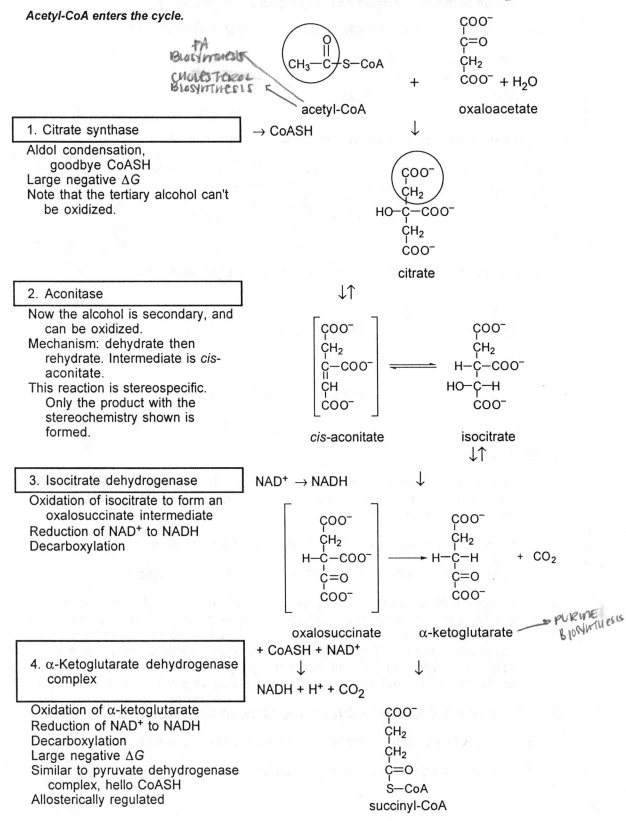

1. Citrate synthase

Aldol condensation,
  goodbye CoASH
Large negative ΔG
Note that the tertiary alcohol can't
  be oxidized.

2. Aconitase

Now the alcohol is secondary, and
  can be oxidized.
Mechanism: dehydrate then
  rehydrate. Intermediate is *cis*-
  aconitate.
This reaction is stereospecific.
  Only the product with the
  stereochemistry shown is
  formed.

3. Isocitrate dehydrogenase

Oxidation of isocitrate to form an
  oxalosuccinate intermediate
Reduction of NAD^+ to NADH
Decarboxylation

4. α-Ketoglutarate dehydrogenase complex

Oxidation of α-ketoglutarate
Reduction of NAD^+ to NADH
Decarboxylation
Large negative ΔG
Similar to pyruvate dehydrogenase
  complex, hello CoASH
Allosterically regulated

108

CITRIC ACID CYCLE, CONTINUED...

REGENERATION OF OXALOACETATE TO CONTINUE THE CYCLE

INHIBITED
W/ ↑ [ATP],
↓ [NAD+],
↓ [FAD],
↓ [ADP]

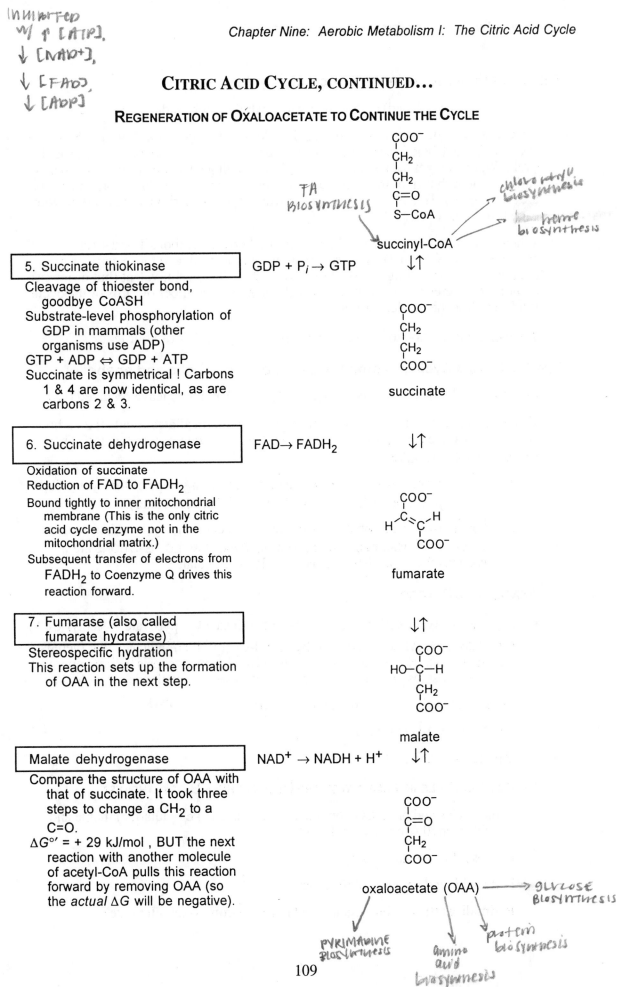

FA
BIOSYNTHESIS

chlorophyll biosynthesis

heme biosynthesis

succinyl-CoA

5. Succinate thiokinase

GDP + P$_i$ → GTP

Cleavage of thioester bond,
goodbye CoASH
Substrate-level phosphorylation of
GDP in mammals (other
organisms use ADP)
GTP + ADP ⇔ GDP + ATP
Succinate is symmetrical ! Carbons
1 & 4 are now identical, as are
carbons 2 & 3.

succinate

6. Succinate dehydrogenase

FAD → FADH$_2$

Oxidation of succinate
Reduction of FAD to FADH$_2$
Bound tightly to inner mitochondrial
membrane (This is the only citric
acid cycle enzyme not in the
mitochondrial matrix.)
Subsequent transfer of electrons from
FADH$_2$ to Coenzyme Q drives this
reaction forward.

fumarate

**7. Fumarase (also called
fumarate hydratase)**

Stereospecific hydration
This reaction sets up the formation
of OAA in the next step.

malate

Malate dehydrogenase

NAD$^+$ → NADH + H$^+$

Compare the structure of OAA with
that of succinate. It took three
steps to change a CH$_2$ to a
C=O.
ΔG°′ = + 29 kJ/mol , BUT the next
reaction with another molecule
of acetyl-CoA pulls this reaction
forward by removing OAA (so
the *actual* ΔG will be negative).

oxaloacetate (OAA) ——→ GLUCOSE
BIOSYNTHESIS

PYRIMIDINE
BIOSYNTHESIS

Amino
acid
biosynthesis

protein
biosynthesis

Fate of Carbon Atoms in the Citric Acid Cycle

To trace specific (labeled) carbon atoms through the citric acid cycle:

Draw out the structures of the citric acid cycle on one page, as in Figure 9.5 of your text. Circle the labeled carbon in each structure, following it around the citric acid cycle. When you get back to citrate, switch to drawing a square around the labeled carbon. For each turn of the cycle, use a different shape (different colored markers are even better). This is a neat way to see where specific carbon atoms end up with each turn of the cycle.

When you got to succinate, you had to choose between carbons 1 and 4 (or between carbons 2 and 3), since succinate is symmetrical, and those carbons are equivalent. Try repeating the exercise above, choosing an equivalent (but different) carbon. It's interesting to see how that will change the number of turns of the cycle it takes to lose the labeled carbon as CO_2.

Try using this method to answer Question 9.5 on page 287 of your text.

The Citric Acid Cycle is Amphibolic (i.e., both catabolic and anabolic).

Catabolic is obvious: 2 carbons enter (acetyl-CoA) and they leave oxidized (CO_2).

To be anabolic, the citric acid cycle must produce intermediates. Wait! If 2 carbons enter and 2 CO_2 leave, how can the citric acid cycle produce intermediates? There are two possible solutions:

1. There must be a way to bypass the CO_2-generating steps of the cycle. This is the glyoxylate cycle, used by plants (and discussed later).

2. There must be additional ways for carbon to enter the cycle (other than via acetyl-CoA). Mammals use this option. ANAPLEROTIC REACTIONS are reactions that replenish citric acid cycle intermediates.

ANAPLEROTIC REACTIONS

- pyruvate → oxaloacetate → citric acid cycle (instead of pyruvate → acetyl-CoA)

  Pyruvate carboxylase is activated by acetyl-CoA (which indicates that there isn't enough OAA). Excess OAA can be used for gluconeogenesis. (Remember that the first two steps of gluconeogenesis are pyruvate → oxaloacetate → PEP.)

- certain fatty acids (also certain amino acids) → succinyl-CoA

- glutamate → α-ketoglutarate

- aspartate → oxaloacetate

CITRIC ACID CYCLE INTERMEDIATES ARE USED IN THESE BIOSYNTHESES:

Amino acids and proteins (from oxaloacetate and α-ketoglutarate); heme and chlorophyll (from succinyl-CoA)

Glucose (from oxaloacetate)

Fatty acids and cholesterol (from acetyl Co-A)

Pyrimidines (from oxaloacetate) and purines (from α-ketoglutarate)

Citric Acid Cycle Regulation

Remember the ultimate goals of the citric acid cycle, and its regulation will make sense. When the cell needs energy or intermediates, the cycle will be activated. The cycle will be inhibited when the cell has plenty of ATP, and its energy needs are being met. It will also be inhibited at low substrate concentrations - including NAD^+, FAD, and ADP.

low $NADH/NAD^+$ ratio[4]	high $NADH/NAD^+$ ratio
low ATP/ADP ratio	high ATP/ADP ratio

Metabolic branch points: Look at Figure 9.10 in your text (or at the list above). The four intermediates mentioned are oxaloacetate, acetyl-CoA, α-ketoglutarate and succinyl-CoA. So, it makes sense that their enzymes are closely regulated. It also makes sense that the two CO_2-generating reactions would be closely regulated.

CITRATE SYNTHASE

$$acetyl\text{-}CoA + oxaloacetate \rightarrow citrate + CoASH$$

Activated by:	Inhibited by:
acetyl-CoA, oxaloacetate, ADP, NAD^+	citrate, succinyl-CoA, ATP, NADH

ISOCITRATE DEHYDROGENASE

$$isocitrate + NAD^+ \rightarrow \alpha\text{-}ketoglutarate + NADH + CO_2$$

Activated by: high ADP, NAD^+	Inhibited by: ATP, NADH

Isocitrate dehydrogenase is also important for citrate metabolism. Only citrate can penetrate the mitochondrial membrane; acetyl-CoA can't. Why would citrate want to leave the mitochondrion? The cell's energy needs are being met, so citrate isn't needed for the citric acid cycle. Citrate can be used to carry acetyl-CoA from the mitochondria to the cytosol, where it can be used in fatty acid synthesis. Citrate activates the first reaction of fatty acid synthesis, and recall that citrate inhibits glycolysis (at PFK-1). Figure 9.12 in your text is a great summary of citrate metabolism.

α-KETOGLUTARATE DEHYDROGENASE

$$\alpha\text{-}ketoglutarate + NAD^+ \rightarrow succinyl\text{-}CoA + NADH + CO_2 + H^+$$

Inhibited by: NADH

Will the citric acid cycle make intermediates or will it make energy? To make intermediates, pyruvate enters the cycle as oxaloacetate (using pyruvate carboxylase). To make energy, pyruvate enters the cycle as acetyl-CoA (using pyruvate dehydrogenase). Obviously, these two enzymes will be closely regulated as well.

Acetyl-CoA activates pyruvate carboxylase and inhibits pyruvate decarboxylase. Why does this make sense? If intermediates are being made, oxaloacetate levels drop and acetyl-CoA will accumulate. Activating pyruvate carboxylase replenishes oxaloacetate; inhibiting pyruvate dehydrogenase prevents more acetyl-CoA from being made.

---

[4] Watch out for the quintessential trick question, which plays with these ratios. Make sure that this makes sense to you, so that you will immediately see that a *low* $NADH/NAD^+$ ratio and a **high** $NAD^+/NADH$ ratio are the same, that they both indicate a need for energy, and that the citric acid cycle will be activated.

THE GLYOXYLATE CYCLE BYPASSES CO_2–GENERATING STEPS

TWO ACETYL-COA REACT TO FORM ONE MALATE

The glyoxylate cycle occurs in plants, some fungi, algae, protozoans, and bacteria.

Glyoxylate enzymes occur in the cytoplasm, except in plants, where they occur in glyoxysomes.

1. A glyoxysome-specific isozyme of citrate synthase

...same as the citric acid cycle...

2. A glyoxysome-specific isozyme of aconitase

...same as the citric acid cycle...

3. Isocitrate lyase

Aldol cleavage

4. Malate synthase

Also requires H_2O

Net:

    Four carbons in,
zero carbons out !!

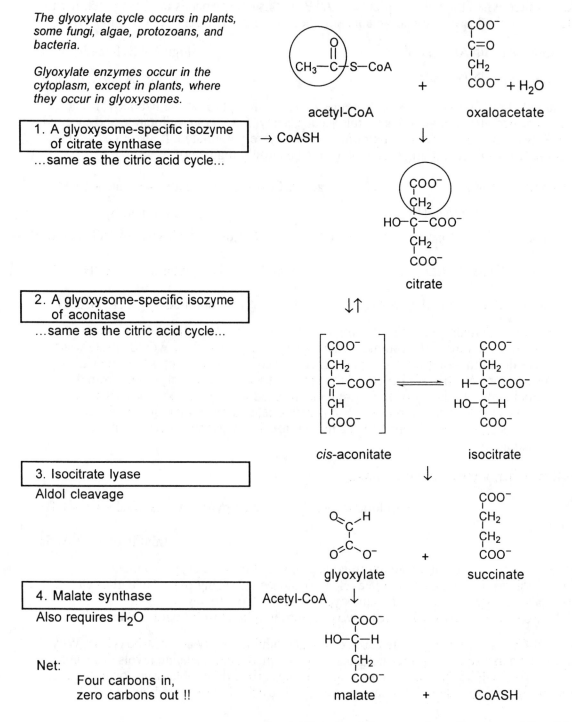

Succinate and malate can each continue around the citric acid cycle to form the intermediate(s) needed

AFTER STUDYING THIS CHAPTER, YOU SHOULD BE ABLE TO HANDLE THESE TYPES OF PROBLEMS AND QUESTIONS:

(ICQ = In-Chapter Question, ICP = In-Chapter Problem, R = Review Question, T = Thought Question)

- Given half-cell potential (E_0') data, calculate $\Delta E_0'$ for a reaction . Determine whether a given redox reaction will proceed in the direction written.   (ICQ-9.1, ICP-9.1)

- Identify redox reactions. Identify oxidizing and reducing agents. Identify whether a molecule is oxidized or reduced in a given reaction.   (ICQ-9.2, ICQ-9.4, R-9)

- Calculate $\Delta G^{\circ\prime}$ using the equation: $\Delta G^{\circ\prime} = -n\, F\, \Delta E_0'$ ($F = 96.5$ kJ/Vmol)
          (ICP-9.1, T-5)

- Identify oxidation states of atoms in molecules.   (ICQ-9.3)

- Trace specific carbons through the pathways learned to date.   (ICQ-9.5, R2)

- Determine whether the citric acid cycle will be activated or inhibited, given one or more of the following data: high/low levels of specific effectors (such as high ATP levels)

     ratios of concentrations of effectors or indicators of energy levels (such as high NAD^+/NADH or low AMP/ATP)

     specific energy conditions (running vs. resting, for example)

          (R4, R8)

CHAPTER 9: ANSWERS TO EVEN-NUMBERED REVIEW QUESTIONS

2. Ancient earth possessed an atmosphere that contained methane, ammonia and was devoid of oxygen. With the development of photosynthesis, oxygen was released into the atmosphere. Subsequently, this oxygen reacted with methane to form carbon dioxide and with ammonia to form molecular nitrogen. The continued release of oxygen produced an oxidizing atmosphere consisting of primarily oxygen, nitrogen, and carbon dioxide.

4. Contracting muscle converts large amounts of ATP to ADP in the muscle cells. The drop in ATP concentration stimulates two key regulatory enzymes of the citric acid cycle. (1) Citrate synthetase catalyzes the condensation of acetyl-CoA and oxaloacetate to give citrate. ATP is an allosteric inhibitor of this enzyme. As concentrations of ATP drop, this enzyme becomes more active. (2) Isocitrate dehydrogenase, which converts isocitrate to α-ketoglutarate is inhibited by high concentrations of ATP and activated by high concentrations of ADP. A reduced ATP concentration also stimulates the conversion of pyruvate to acetyl-CoA catalyzed by pyruvate dehydrogenase.

6. Citrate, produced in the mitochondria by the citric acid cycle, is transported across the mitochondrial membrane into the cytoplasm. Once in the cytoplasm, citrate is cleaved by citrate lyase to acetyl-CoA and oxaloacetate. The acetyl-CoA is then used to synthesize fatty acids as well as other biomolecules. The oxaloacetate is reduced to malate and moved across the mitochondrial membrane where it is reoxidized to oxaloacetate, the citric acid cycle intermediate.

8. A high NADH/NAD⁺ ratio (a) indicates that a cell's energy requirements are currently being met. NADH inhibits pyruvate dehydrogenase, citrate synthase, isocitrate dehydrogenase and α-ketoglutarate dehydrogenase. A high ADP/ATP ratio (b) and low citrate concentration (d) are indicators of a low cell energy state. A high ADP/ATP ratio (b) stimulates isocitrate dehydrogenase, citrate synthase, ATP synthase and oxidative phosphorylation. A high acetyl-CoA concentration (c) inhibits pyruvate dehydrogenase and stimulates pyruvate carboxylase and fatty acid synthesis. A low citrate concentration (d) stimulates citrate synthase, depresses fatty acid synthesis, and stimulates PFK-1. A high succinyl-CoA concentration (e) inhibits citrate synthase and α-ketoglutarate dehydrogenase.

CHAPTER 9: ANSWERS TO EVEN-NUMBERED THOUGHT QUESTIONS

2. After fructose is absorbed from the intestine it is transported to liver where it is converted to fructose-1-phosphate, which then enters the glycolytic cycle when it is cleaved to form DHAP and glyceraldehyde. Its carbon atoms then are transformed into pyruvate. Pyruvate is converted to acetyl-CoA which then can enter the citric acid cycle. Fructose that is taken up by adipose and muscle tissue enters the glycolytic pathway by being converted to fructose-6-phosphate by hexokinase

4. Glutamate is converted to α-ketoglutarate in a transamination reaction. The α-ketoglutarate enters the citric acid cycle and is eventually converted to CO_2 and H_2O.

Aerobic Metabolism II:
Electron Transport and
Oxidative Phosphorylation

ABBREVIATIONS AND ACRONYMS

cyt   cytochrome

ETC   electron transport chain

FAD   flavin adenine dinucleotide

FMN   flavin mononucleotide

NAD   nicotinamide adenine dinucleotide

ROS   reactive oxygen species

SOD   superoxide dismutase

UQ   ubiquinone, or coenzyme Q

ELECTRON TRANSPORT

WHY BOTHER?   *Energy.* Aerobic respiration (using oxygen to generate energy from food) yields much more energy from nutrients than does fermentation. This energy is used to do work, including synthesizing ATP, pumping Ca^{2+} into the mitochondrial matrix, and generating heat in brown adipose tissue. ATP hydrolysis is used to fuel endergonic reactions, run molecular machines, and in general, to act as the energy "currency" for the cell.

O_2 PROPERTIES THAT ARE RELEVANT TO AEROBIC METABOLISM:

O_2 is everywhere (almost).

O_2 diffuses easily across cell membranes

O_2 is highly reactive so it accepts electrons readily. (That could also be bad - O_2 can also form ROS.)

ELECTRON DONORS: NADH and $FADH_2$ (from glycolysis, citric acid cycle, and fatty acid oxidation)

INTERMEDIARIES THAT CAN TRANSFER 1 ELECTRON AT A TIME:

FMN, FAD, and UQ have resonance-stabilized radicals

Iron-Sulfur Clusters and Cytochromes contain iron (and/or copper)

FMN – FLAVIN MONONUCLEOTIDE

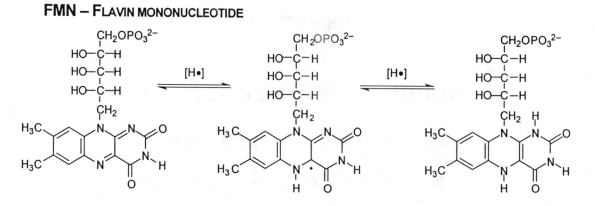

FMN
(oxidized or quinone form)

FMNH•
(radical or semiquinone form)

FMNH₂
(reduced or hydroquinone form)

The structure of FAD is FMN with another phosphate, another sugar, and an adenine. Since FMN and FAD have the same three-ring "isoalloxazine group" that stabilizes a radical intermediate, they can both transfer one electron at a time.

UBIQUINONE (UQ) (OR, COENZYME Q)

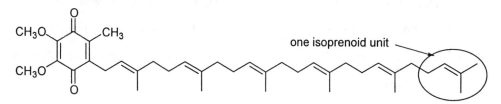

one isoprenoid unit

Mammalian UQ has ten isoprenoid units; some bacteria have six, as shown above. The long hydrocarbon tail makes UQ able to diffuse within the inner membrane, so it's mobile - it can travel to carry electrons between the donors at complex I or II and the acceptor at complex III. The semiquinone intermediate is a radical that's stabilized by resonance.

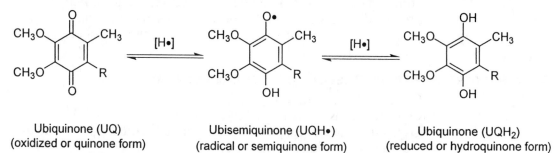

Ubiquinone (UQ)
(oxidized or quinone form)

Ubisemiquinone (UQH•)
(radical or semiquinone form)

Ubiquinone (UQH₂)
(reduced or hydroquinone form)

CYTOCHROMES: A SERIES OF ELECTRON TRANSPORT PROTEINS THAT CARRY AN FE-CONTAINING HEME PROSTHETIC GROUP (SIMILAR TO THE HEME IN HEMOGLOBIN)

Example: cyt c (Fe^{3+}) + 1 e^- → cyt c (Fe^{2+})
oxidized form   reduced form

ELECTRON TRANSPORT AND ITS COMPONENTS

LOCATION: Eukaryotes: in the inner membrane of the mitochondria
              Aerobic prokaryotes: in the plasma membrane

Complex and its Common Name Coenzymes; Components	Path of Electrons	
	From Donor:	To Receiver:
Complex I NADH Dehydrogenase Complex: FMN, ~ 7 Fe-S centers, 25 different polypeptides results in 4 H$^+$ transported to intermembrane space	from NADH	to FMN → FeS → FeS → UQ → UQH$_2$
Complex II Succinate Dehydrogenase Complex: Succinate Dehydrogenase, FAD (bound), and 2 Fe-S proteins	from succinate	to FAD → FeS → UQ→ UQH$_2$
In some cell types: Glycerol-3-phosphate dehydrogenase (on outer face of inner mito. membrane)	cytoplasmic NADH →	to DHAP (forms glycerol-3P) → FAD → FeS → UQ→ UQH$_2$
Acyl-CoA dehydrogenase (matrix side of inner membrane) (first step of fatty acid oxidation)	from fatty acyl-CoA	to FAD → FeS centers → UQ→ UQH$_2$
Complex III Cytochrome bc$_1$ Complex 1 Fe-S center cyt b$_L$ cyt b$_H$ cyt c$_1$ cyt c: loosely attached to inner membrane -space side results in 4 H$^+$ transported to intermembrane space	from UQH$_2$ (space-side) UQH• (space-side, from UQH$_2$ above) from UQH$_2$ (space-side)	1 e$^-$ to Fe-S → cyt c$_1$ → cyt c 1 e$^-$ to cyt b$_L$ → cyt b$_H$ → UQ (forms UQH•, matrix-side) 1 e$^-$ to cyt b$_L$ → cyt b$_H$ → UQH•, matrix-side (forms UQH$_2$)
Complex IV Cytochrome Oxidase cyt a, cyt a$_3$ 2 Cu: Cu$_A$ near cyt a; Cu$_B$ near cyt a$_3$ results in 2 H$^+$ (per cyt c) transported to intermembrane space	from cyt c 1 e$^-$ at a time	to Cu$_A$ → cyt a → cyt a$_3$ → Cu$_B$ → O$_2$ (forms H$_2$O) O$_2$ + 4 H$^+$ + 4e$^-$ → 2 H$_2$O

Electron Transport Inhibitors

(Inhibitors have been used to elucidate the order of the ETC components.)

Complex I: amytal, rotenone   Complex III (cyt b):   antimycin A

Complex IV: CO, N_3^-, CN^- (carbon monoxide, azide, cyanide)

OXIDATIVE PHOSPHORYLATION = USING ETC ENERGY TO MAKE ATP

The Chemiosmotic Theory

H^+ ions are transported from the matrix to the intermembrane space during ETC electron transfer. This creates a proton gradient (the protonmotive force, Δp) across the inner membrane.

The H^+ ions in the intermembrane space can only flow back to the matrix (down the proton gradient) through special channels. As they pass through the channels, ATP synthesis occurs. Yay! ☺

Evidence for the chemiosmotic theory:

1. Working mitochondria take in O_2 and expel H^+.
2. If the mitochondrial membrane is disrupted, the ETC continues but ATP synthesis stops.
3. Uncouplers and ionophores allow H^+ to leak across the membrane (they're kind of like a free route around a toll booth). Uncouplers carry H^+ across, and ionophores form channels through the membrane. Examples: dinitrophenol (uncoupler); gramicidin A (ionophore)

ATP Synthesis by ATP Synthase (in the mitochondrial inner membrane)

ATP synthase resembles a lollipop, with the stick in the mitochondrial inner membrane and the lollipop facing the matrix.

ATP Synthase Components:

- F_1 unit (active ATPase) with three nucleotide-binding sites; contains five subunits: 3 $\alpha\beta$ dimers, γ, δ, and ϵ; on the inner membrane's matrix side

- F_0 unit - transmembrane channel for H^+, contains three subunits (in the ratio a, b_2, and c_{12})

Each H^+ that passes through turns the rotor (ϵ, γ, c_{12}) by 120° while the stator (a, b_2, δ, α_3, β_3) stays still. After three H^+ ions pass through, a full turn is made and an ATP is released.

Rotation changes conformation of the protein, which changes the binding affinity.
Conformation A: binds ADP and Pi weakly
Conformation B: brings substrates closer, facilitates ATP formation
Conformation C: nonbinding: expels ATP

Control of Oxidative Phosphorylation

P/O ratio: moles P_i used (to make ATP) per O_2 reduced to H_2O; This ratio correlates the ETC with ATP synthesis. The max P/O ratio measured for NADH is 2.5 and for $FADH_2$ is 1.5.

Respiratory control: control of aerobic respiration by ADP (that is, there has to be enough ADP and P_i available to be able to make ATP)

ATP synthase: inhibited by ATP, activated by ADP and P_i

Ratio of ATP and ADP controlled by:

ADP-ATP translocator: ships ATP out (of the matrix) and imports ADP (into the matrix from the intermembrane space). Remember that the intermembrane space is more positive (because of all of those H^+ ions) and the matrix is more negative. So, shipping out ATP (with its extra negative charge) is favored.

Phosphate translocase: imports $H_2PO_4^-$ (into the matrix) with an H^+ from the intermembrane space. (This is driven by the proton gradient.)

So, for each ATP synthesized, FOUR H^+ total must pass through the membrane: three to turn the ATP synthase rotor and one to supply the phosphate.

For the Complete Oxidation of Glucose:

The electrons captured by NADH (from glycolysis in the cytoplasm) has to be transported into the mitochondrion, since NADH can't get through the membrane.

GLYCEROL PHOSPHATE SHUTTLE

Converts cytoplasmic NADH (from glycolysis) to inner membrane $FADH_2$

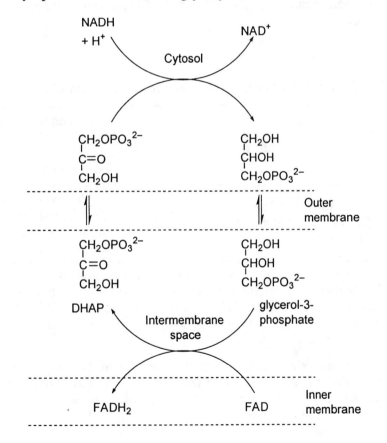

MALATE-ASPARTATE SHUTTLE

Transports NADH and an H^+ from the intermembrane space into the matrix so it can enter the ETC. See Figure 10.17b, page 218 of your text. Because of that H^+ import, the number of ATP made per NADH is a bit less.

ONE MORE SHUTTLE

Transport of two citric-acid-cycle-produced ATPs from the matrix to the cytoplasm (so they can be used) cost 2 H^+ (total), so that also reduces the amount of ATP that can be produced from glucose oxidation.

The Complete Oxidation of Glucose:

$$C_6H_{12}O_6 + 6\ O_2 + 30\ ADP + 30\ P_i \rightarrow 6\ CO_2 + 6\ H_2O + 30\ ATP$$

Uncoupled Electron Transport and Heat Generation

Sometimes a little leak is a good thing. When uncouplers and ionophores allow H^+ ions to cross the membrane, the energy that would have been used to make ATP is released as heat. That's nonshivering thermogenesis: "producing heat without shivering." Brown fat cells have many mitochondria that literally are little furnaces. In their inner membranes, they contain thermogenin (uncoupling protein or UCP).

Norepinephrine is a neurotransmitter (released from neurons that end in brown fat) that ultimately causes fat molecules to hydrolyze. The fatty acids produced activate thermogenin.

OXIDATIVE STRESS

Reactive Oxygen Species (ROS) are formed from O_2

$O_2\overset{-}{\bullet}$	superoxide radical	H_2O_2	hydrogen peroxide, H-O–O-H (can cross membranes) $Fe^{2+} + H_2O_2 \rightarrow Fe^{3+} + \bullet OH + OH^-$
$\bullet OH$	hydroxyl radical - very reactive, can initiate a chain reaction	$^1O_2$	singlet oxygen (an unpaired electron absorbs energy and shifts to a higher orbital, i.e., an excited state); formed from superoxide or peroxide

If you've never seen these radicals before, it's because they are so reactive you can't buy them. Hydrogen peroxide isn't a radical - yet - but its O–O single bond can react to form radicals. (That's why drugstore peroxide is in a brown bottle - light energy would degrade it.) ROS formation is linked to aging, radiation, smoking, and certain diseases.

ROS react as soon as they're formed, oxidizing and damaging whatever they touch: inactivating enzymes, depolymerizing polysaccharides, breaking DNA, destroying membranes, etc. Macrophages and neutrophils seek out microorganisms and damaged cells and destroy them with a respiratory burst - ROS that they've made for that purpose.

Antioxidant Enzyme Systems Destroy ROS

SUPEROXIDE DISMUTASE (SOD): $2\,O_2^{\bullet-} + 2\,H^+ \rightarrow H_2O_2 + O_2$

Two types in humans: Cu-Zn isozyme in the cytoplasm, Mn isozyme in the mitochondrial matrix.

Lou Gehrig's disease is caused by a mutation in the gene that codes for cytosolic Cu-Zn isozyme of SOD.

GLUTATHIONE PEROXIDASE: $GS\text{–}H \rightarrow GS\text{–}SG$

Contains Se

$$2\,G\text{–}S\text{–}H + R\text{–}O\text{–}O\text{–}H \rightarrow G\text{–}S\text{–}S\text{–}G + R\text{–}O\text{–}H + H_2O$$

(If R = H, then ROOH is simply hydrogen peroxide, which forms water.)

Glutathione reductase regenerates GSH:

$$G\text{–}S\text{–}S\text{–}G + NADPH + H^+ \rightarrow 2\,GSH + NADP^+$$

NADPH is from: pentose phosphate pathway; isocitrate dehydrogenase; malic enzyme

CATALASE (CONTAINS HEME) PUTS H_2O_2 TO WORK TO OXIDIZE COMPOUNDS

$$RH_2 + H_2O_2 \rightarrow R + 2\,H_2O \qquad\qquad \text{Also: } 2\,H_2O_2 \rightarrow 2\,H_2O + O_2$$

Antioxidant Molecules

Antioxidant molecules can accept an electron from ROS to form a radical that is stabilized by resonance and won't be as reactive. Great examples of the stabilized radicals are FMNH• and UQH• (see these radical structures in an earlier section in this chapter). Vitamins E and C are further examples of antioxidants.

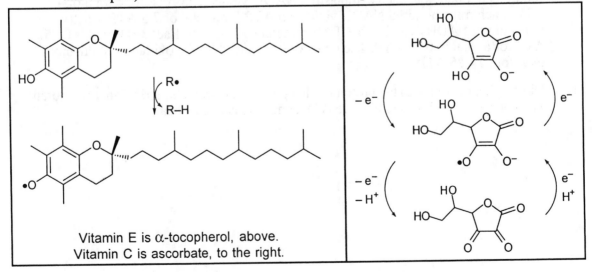

Vitamin E is α-tocopherol, above.
Vitamin C is ascorbate, to the right.

CHAPTER 10: ANSWERS TO EVEN-NUMBERED REVIEW QUESTIONS

2. Principal sources of electrons for the mitochondrial electron transport system are NADH and $FADH_2$.

4. The chemiosmotic coupling theory has the following principal features: As electrons pass through the electron transport chain (ETC), protons are transported from the matrix and released into the inner membrane space. As a result, an electrical potential (Ψ) and a proton gradient (ΔpH) are created across the inner membrane. The electrochemical proton gradient is sometimes referred to as the proton motive force (Δp). Protons, which are present in the intermembrane space in great excess, can pass through the inner membrane and back into the matrix down their concentration gradient only through the proton-translocating ATP synthase.

6. ATP synthesis requires the presence of a proton gradient across the mitochondrial membrane. Dinitrophenol, which has an ionizable hydrogen, can diffuse across the mitochondrial membrane. As it diffuses across the membrane, protons are transported from one side of the mitochondrial membrane to the other, thereby disrupting the proton gradient and interfering with ATP synthesis.

8. Oxygen is widely used as an energy source because it makes possible greater energy yields from food molecules and it is readily available.

10. ROS damage cells by inactivating enzymes, depolymerizing polysaccharides, breaking DNA, and destroying membranes.

12. A genetic defect with survival value is G-6-PD deficiency. The lower NADPH and GSH concentrations in the red blood cells of G-6-PD-deficient individuals inhibit the growth of Plasmodium (the malarial parasite), a major killer of humans.

CHAPTER 10: ANSWERS TO EVEN-NUMBERED THOUGHT QUESTIONS

2. The oxidation of one mole of ethanol to acetyl-CoA produces two moles of NADH. The conversion of acetate to carbon dioxide and water through the citric acid cycle produces 3 NADH, 1 $FADH_2$, and 1 GTP. Assuming that the aspartate-malate shuttle is in operation, each cytoplasmic NADH yields 2.25 ATP for a total of 4.5 ATP. Each mitochondrial NADH yields 2.5 ATP for a total of 7.5 ATP. Each molecule of $FADH_2$ yields 1.5 ATP for a total of 1.5 ATP. Each GTP yields 0.75 ATP for a total of 0.75 ATP. The total ATP produced by the oxidation of ethanol is therefore 14.25 ATP.

4. Dinitrophenol collapses the proton gradient across the mitochondrial inner membrane. The energy normally used to drive ATP synthesis is lost as heat.

CHAPTER ELEVEN

Lipids and Membranes

LIPID CLASSES

Fatty Acids and their Derivatives

Fatty acids have a long hydrocarbon chain (12-20 carbons) and one carboxylic acid. They are components of lipid molecules, primarily triacylglycerols and membrane-bound lipids. Fatty acids are amphiphilic, meaning they contain both hydrophobic and hydrophilic ends.

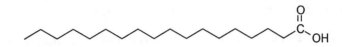

SATURATED VS. UNSATURATED (*CIS* VS. *TRANS*)

If the hydrocarbon chain contains at least one double bond, the fatty acid is *unsaturated* (not completely filled with hydrogens). If the hydrocarbon chain contains no double bonds, it's *saturated* (completely filled with hydrogens). Monounsaturated fatty acids have one double bond, and polyunsaturated fatty acids have two or more double bonds. Double bonds can be oxidized.

MELTING POINTS AND FLUIDITY OF A FATTY ACID:

The lower the melting point, the more fluid a fatty acid (or lipid) is likely to be at a given temperature. What determines whether the fatty acid is a solid or liquid at a given temperature? 1) number of carbons 2) number and type of double bonds.

The longer the hydrocarbon chain, the higher the melting point (that is, it takes more energy to put those long chains in motion).

The greater the number of double bonds, the lower the melting point. A *cis* double bond causes a "kink" in the chain that prevents *cis*-unsaturated fatty acids from packing as closely as saturated fatty acids do. So, it takes less energy to melt *cis*-unsaturated fatty acids (than a saturated fatty acid with the same number of carbons), and they have lower melt points. (*Trans*-unsaturated fatty acids are similar in structure to saturated fatty acids.)

Example: Compare the melting points of these fatty acids. Why does stearic acid have the highest melting point, and linoleic acid have the lowest?

myristic acid (14 carbons)	$CH_3(CH_2)_{12}COOH$	54°C
palmitic acid (16 carbons)	$CH_3(CH_2)_{14}COOH$	63°C
stearic acid (18 carbons)	$CH_3(CH_2)_{16}COOH$	70°C
oleic acid	$CH_3(CH_2)_7CH=CH(CH_2)_7COOH$	4°C
linoleic acid	$CH_3(CH_2)_4CH=CHCH_2CH=CH(CH_2)_7COOH$	−12°C

Naturally occurring fatty acids are typically:

unbranched, even number of carbon atoms, if unsaturated: *cis* form

DESIGNATION:

(total number of carbons) : (number of double bonds)$^{\Delta(\text{location of double bond})}$

Location of double bond = counting from the carboxyl group, the first carbon that begins the double bond

Example:
Linoleic acid, $18:2^{\Delta 9, 12}$, has 18 total carbons and two double bonds, one located between carbons 9-10 and the second located between carbons 12-13.

NONESSENTIAL VS. ESSENTIAL FATTY ACIDS

Nonessential fatty acids - can be synthesized by the body

Essential fatty acids - must be obtained from the diet (sources include some vegetable oils, nuts, seeds); examples: linoleic and linolenic acids; metabolic precursors; symptoms of dietary deficiency include: dermatitis, poor wound healing, reduced resistance to infection, alopecia, thrombocytopenia

CHEMICAL REACTIONS OF FATTY ACIDS

- carboxylic acid + alcohol = ester (reversible)

- hydrogenation of unsaturated fatty acids = saturated fatty acids

- oxidative attack of unsaturated fatty acids (Chapter 12)

- protein acylation - fatty acid groups (acyl groups) covalently attach to proteins

    example of fatty acid reactions in the body: transport of fatty acids: esterification to serum proteins allows transport from fat cells to body cells; acyl transfer reactions allow fatty acids to enter cells

Triacylglycerols - esters of glycerol with three fatty acids

All three hydroxyl groups of glycerol have a fatty acid attached via an ester bond. For each ester bond that's formed, a water molecule is lost.

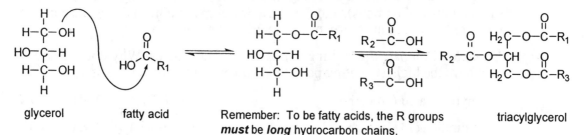

glycerol   fatty acid   Remember: To be fatty acids, the R groups   triacylglycerol
                                      must be ***long*** hydrocarbon chains.

FATS VS. OILS: AT ROOM TEMPERATURE, FATS ARE SOLID, OILS ARE LIQUID. WHY?

Fats have more saturated fatty acids, and oils have more unsaturated fatty acids. Partial hydrogenation: commercial hydrogenation of double bonds in oils converts them to fats

FUNCTIONS OF TRIACYLGLYCEROLS:

To store and transport fatty acids

Triacylglycerols store energy more efficiently than glycogen because:

1. hydrophobic; coalesce into compact anhydrous droplets in adipocytes; take up 1/8 the volume of glycogen (because glycogen binds a substantial amount of water)

2. when degraded (oxidized), triacylglycerols release more energy per gram than glycogen

Insulation; poor conductor of heat, prevents heat loss

Water-repellent - secreted by specialized glands to make fur or feathers water-repellent

Plants - important energy reserve in fruits (avocados and olives) and seeds (peanut, corn, palm, safflower, soybean)

Soapmaking (saponification): soap = sodium or potassium salts of fatty acids; emulsifying agents; how soap works

Wax Esters

Complex mixtures - primarily esters from long-chain fatty acids and long-chain alcohols, also contain hydrocarbons, alcohols, fatty acids, aldehydes, sterols

Protective coatings on leaves, stems, fruits and the skin and fur of animals

Examples: carnauba wax (melissyl cerotate - 26-carbon carboxylic acid esterified to a 30-carbon alcohol) and beeswax (triacontyl hexadecanoate)

Phospholipids: phosphoglycerides and sphingomyelins

FUNCTIONS: Structural components of membranes
Emulsifying agents
Surface active agents (lower the surface tension of a liquid so it can spread out over a surface)

AMPHIPATHIC: hydrophilic (polar head group) and hydrophobic (hydrocarbon chain)
In water, phospholipids spontaneously rearrange into micelles and/or bilayers.

TYPES: Phosphoglycerides: glycerol, fatty acids, phosphate, and an alcohol
Sphingomyelins: sphingosine, fatty acids, phosphate, and an alcohol

PHOSPHOGLYCERIDES:

The simplest phospholipid is phosphatidic acid: glycerol-3-phosphate esterified to two fatty acids, typically with 16-20 carbons each and: • R_1 = saturated fatty acid esterified to C-1 • R_2 = unsaturated fatty acid esterified to C-2	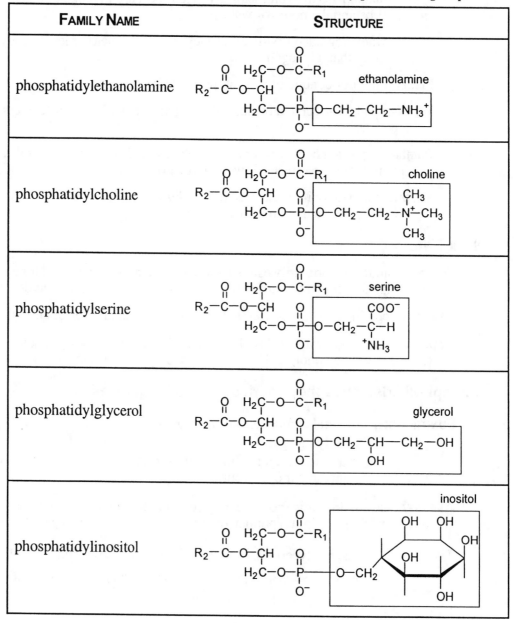Phosphatidic acid

Other types of molecules can be attached to the phosphate group via a phosphodiester bond linkage. Note how each has a very polar *head group*.

FAMILY NAME	STRUCTURE
phosphatidylethanolamine	ethanolamine
phosphatidylcholine	choline
phosphatidylserine	serine
phosphatidylglycerol	glycerol
phosphatidylinositol	inositol

Another function of phosphoglycerides: intracellular signal transduction. *Example*: phosphatidylinositol-4,5-bisphosphate (PIP$_2$) phosphatidylinositol cycle

Sphingolipids: Another lipid component of biological membranes

Sphingosine is the "backbone" of sphingolipids (in animals; phytosphingosine is in plants).	H—C=CH—(CH$_2$)$_{12}$—CH$_3$ H—C—OH H—C—NH$_3^+$ CH$_2$OH Sphingosine
Ceramides have a fatty acid attached to the amino group of sphingosine.	H—C=CH—(CH$_2$)$_{12}$—CH$_3$ H—C—OH O H—C—NH—C—R Ceramide CH$_2$OH
The last –OH group of ceramide can also have substituents attached to it. Sphingomyelin, an important brain lipid, has a phosphorylcholine or phosphorylethanolamine at the last position. Sphingomyelin is found in myelin sheath of nerve cells (and in other cell membranes). It insulates nerve cells to allow for rapid transmission of nerve impulses.	H—C=CH—(CH$_2$)$_{12}$—CH$_3$ H—C—OH O H—C—NH—C—R CH$_2$ O CH$_3$ O—P—O—CH$_2$—CH$_2$—N$^+$—CH$_3$ O$^-$ CH$_3$ Sphingomyelin

Sugars can also be found at the terminal hydroxyl of ceramide.

Glycolipids (sugar + lipid) (also glycosphingolipids) - monosaccharide, disaccharide, or oligosaccharide O-glycosidic linkage to ceramide; contain no phosphate

Cerebrosides - monosaccharide head group (nonionic)

Galactocerebrosides - found in cell membranes of the brain

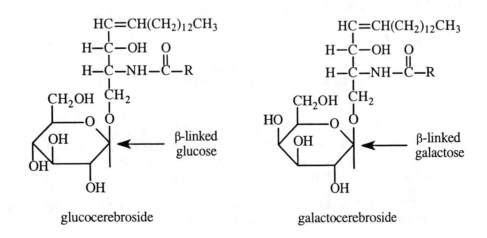

glucocerebroside   galactocerebroside

Sulfatides - sulfated cerebroside, negatively charged

Gangliosides - sphingolipids with oligosaccharide groups with at least one sialic acid residue; Names: subscript letter (M, D, or T = 1, 2, or 3 sialic acid residues) and numbers (the sequence of sugars that are attached to ceramide)

Isoprenoids: Terpenes and Steroids

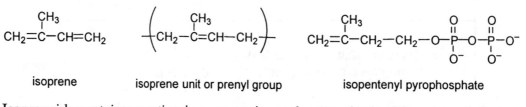

isoprene   isoprene unit or prenyl group   isopentenyl pyrophosphate

Isoprenoids contain repeating isoprene units, and are synthesized from acetyl-CoA (not isoprene), beginning with the formation of isopentenyl pyrophosphate.

TERPENES - classified according to number of isoprene residues; found in essential oils of plants.

# isoprene units	Type
2....	monoterpene
3....	sesquiterpene
4....	diterpene
6....	triterpene
8....	tetraterpene (examples: carotenes, xanthophylls)

MIXED TERPENOIDS - nonterpene components attached to isoprenoid groups (prenyl or isoprenyl groups) *Examples*: vitamins E and K, ubiquinone, some cytokinins

PROTEIN PRENYLATION - prenyl groups covalently attached to proteins; function isn't clear; typically farnesyl and geranylgeranyl groups

STEROIDS - derivatives of triterpenes, found in all eukaryotes, few bacteria

STEROID STRUCTURE:

4 fused rings; may contain double bonds and various substituents (OH, C=O, alkyl groups); sterols contain an OH

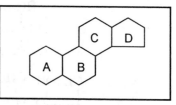

Examples: Cholesterol: functions as an animal cell membrane component and as a precursor to steroid hormones, vitamin D and bile salts (emulsifying agents that aid in the digestion of fats in the small intestine); cholesterol usually stored within cells as a fatty acid ester

Many hormones and vitamins are steroids. Testosterone, progesterone, and cortisol are examples of steroid hormones.

Example of a steroid derivative: cardiac glycosides - increase the force of cardiac muscle contraction; digitoxin - a glycoside in digitalis, an extract from the foxglove plant; used to treat congestive heart failure; toxic in higher than therapeutic doses; inhibits Na^+-K^+ATPase.

Lipoproteins transport lipid molecules through the bloodstream

Molecular complexes found in mammalian blood plasma; also contains lipid-soluble antioxidants (α-tocopherol, carotenoids)

APOLIPOPROTEINS (or apoproteins) are the protein components of lipoproteins (that is, the lipid portion is *A*bsent).

CLASSIFIED ACCORDING TO DENSITY:

higher lipid content = less dense

higher protein content = more dense

LIPOPROTEIN	FUNCTION
Chylomicrons large, extremely low density	Transports dietary triacylglycerols and cholesteryl esters from intestine to muscle and adipose tissues
Very Low Density Lipoproteins (VLDL) synthesized in the liver	Transports lipids to tissues
Low-Density Lipoproteins (LDL) converted from triacylglycerol-depleted VLDLs	Transports cholesterol to tissues; engulfed by cells after binding to LDL receptors
High-Density Lipoprotein (HDL) produced in the liver	Scavenges excess cholesterol from cell membranes: A plasma enzyme transfers a fatty acid residue from lecithin to cholesterol, forming a cholesteryl ester that is transported by HDL to the liver, which converts most of it to bile acids

Lipoproteins and Atherosclerosis

Atherosclerosis - atheromas, or plaque, accumulate on the inside of arteries

Foam cells - macrophages that appear foam-like due to filling with lipid from LDL deposits (mainly cholesterol and cholesteryl esters); have LDL receptors; LDL binds to receptors and initiates endocytosis; ingest LDL

Coronary artery disease - heart muscle deprivation of oxygen and nutrients because of plaque build-up on arterial walls

High plasma LDL levels directly correlated with high risk for coronary artery disease

FURTHER NOTES ON CHOLESTEROL:[1]

1. Cholesterol is required for survival because of its essential function in cell membranes and because it's the precursor of most steroid hormones;

2. The LDL receptor is unique in its regulation because it down-regulates itself; thus, the more LDL we have, the less will be taken in by the cells and as a conclusion an individual will have more circulating LDL. Most receptors upregulate their synthesis according to the concentration of their ligand. This is not the case for the LDL receptor. Once the need of the cell for cholesterol is satisfied, the cell shuts-down its production for LDL receptors. LDL particles are then attacked by the macrophages. The plaque on the inside of the arteries is built because macrophages burst because of the excess of LDL, and damage the endothelium. The damage initiates a coagulation response which, in turn, will create a cell plug to stop the blood from leaking outside the vasculature. A continuous coagulation reaction on the same spot would create a bigger and bigger plug, which will end up obstructing the vessel.

MEMBRANES

Membrane Structure:

FLUID MOSAIC MODEL - proteins float within a lipid bilayer; the types and proportion of proteins and lipids depend on the type of cell (or organelle).

MEMBRANE LIPIDS

1. Membrane fluidity - determined by percentage of unsaturated fatty acids in phospholipids; more unsaturated chains = more fluid

   Cholesterol moderates fluidity because it has both rigid and flexible components

   Lateral diffusion (rapid) vs. flip-flop (rare)

2. Selective permeability: A bilayer is a good barrier to charged molecules like Na^+, K^+, Cl^-, and most polar molecules. Nonpolar substances diffuse through the lipid bilayer down their concentration gradients. Transport across the membrane is controlled by carrier proteins or protein channels.

3. Self-sealing capability - (Figure 11.22) - spontaneous and immediate

4. Asymmetry: different lipid components facing inside vs. facing outside

MEMBRANE PROTEINS

Classified by function: structural, enzymes, hormone receptors, transport

Classified by structural relationship to membrane: integral vs. peripheral

Integral proteins in red blood cell membrane:

---

[1] Thanks to Dr. Michael Kalafatis for these comments on cholesterol.

Glycophorin - 60% carbohydrate (by weight) includes the ABO and MN blood group antigens, function is unknown

Anion channel protein (band 3) - CO_2 transport in blood; HCO_3^- exchanged with Cl^- (chloride shift)

Peripheral proteins in red blood cell membrane: spectrin, ankyrin, band 4.1

Preserves biconcave cell shape - allows rapid diffusion of O_2

Links cytoskeleton and membrane

Anion channel protein linked to ankyrin linked to spectrin linked to band 4.1 linked to actin filaments (a cytoskeleton component); band 4.1 also binds to glycophorin

Membrane Function

MEMBRANE TRANSPORT OF NUTRIENTS, WASTES, IONS, ETC.

PASSIVE TRANSPORT - no energy input needed; transport down a concentration gradient

SIMPLE DIFFUSION - spontaneous transport down a concentration gradient

The greater the gradient, the faster the rate. (example: diffusion of gases such as O_2 and CO_2)

FACILITATED DIFFUSION - special channels or carriers increase the rate at which certain solutes (large or charged molecules) move down their concentration gradients

CHANNELS - tunnel-like transmembrane proteins, designed for a specific solute, often chemically or voltage-regulated

Chemically regulated channels - open or close in response to a specific chemical signal (example: chemically gated Na^+ channel)

Voltage-regulated channels

Membrane potential - electrical gradient across a membrane: decrease = depolarization; reestablishment = repolarization (example: voltage-gated Na^+ or K^+ channel)

CARRIERS OR PASSIVE TRANSPORTERS - solute binds to carrier on one side, causing a conformational change in the carrier that results in translocation across the membrane, carrier releases solute (example: red blood cell glucose transporter)

ACTIVE TRANSPORT needs energy to transport against a concentration gradient

PRIMARY ACTIVE TRANSPORT - ATP hydrolysis provides the energy ; Example: Na^+-K^+ pump (or Na^+-K^+ ATPase pump)

SECONDARY ACTIVE TRANSPORT - primary active transport generates a concentration gradient that is used to move substances across membranes

IMPAIRED MEMBRANE TRANSPORT MECHANISMS

Example: Cystic fibrosis (CF) is caused by a missing or defective plasma membrane glycoprotein (cystic fibrosis transmembrane conductance regulator (CFTR)) - a chloride channel in epithelial cells, failure results in retention of Cl⁻ within cells

UNIPORTERS, SYMPORTERS, ANTIPORTERS

MEMBRANE RECEPTORS: EXAMPLE: LDL RECEPTOR-MEDIATED ENDOCYTOSIS

AFTER STUDYING THIS CHAPTER, YOU SHOULD BE ABLE TO SOLVE THESE TYPES OF PROBLEMS:

(R = Review Question, T = Thought Question)

- Predict relative melt points of fatty acids, given their names or structures

- Membrane fluidity (R6, T6, T7)

- Draw a fatty acid given its designation, for example: $18:3^{\Delta 9,\ 12,\ 15}$

- Draw a triacylglycerol, given the fatty acids' names, structures, or designations

- Draw a phospholipid or sphingomyelin, given the fatty acids and the alcohol (R4)

- Determine the terpene class; locate isoprene units in terpenes (R13)

- Given the structure of a lipid, determine its class (R3)

- Transport across a membrane: Differentiate between types of transport. Identify the type of transport given the molecule to be transported. (R8, R10, R18, R19, R20, R21)

- Differentiate between types and functions of lipoproteins (R5, R16, T3)

CHAPTER 11: ANSWERS TO EVEN-NUMBERED REVIEW QUESTIONS

2. a. Phospholipids perform major roles as structural components of membranes, emulsifiers and surface active agents.

    b. Plant and animal membranes contain large amounts of sphingolipids.

    c. Oils serve as an important energy reserve of fruits and seeds.

    d. Waxes serve as protective coatings on the surface of leaves and stems, on the fur of animals and on the shells of insects.

    e. Steroids play an important structural role in animal membranes. Certain steroids act as hormones.

    f. By acting as light trapping pigments, carotenoids play an important role in photosynthesis.

4. The polar head region containing the ester and amide linkages would be hydrophilic; the hydrocarbon tail region would be hydrophobic.

6. Unsaturated fatty acid content increases fluidity, whereas cholesterol decreases fluidity.

8. Both a and c are true. (Ionophores are discussed on page 310.)

10. When acetylcholine binds to the acetylcholine receptor complex in muscle cell plasma membrane, sodium ions flow into the muscle cell and a smaller number of potassium ions flow out. During the repolarization phase of muscle contraction, potassium ions flow out of the cell through voltage-regulated potassium channels.

12. Prostaglandins have major recognized roles in all of the following processes: reproduction, digestion, respiration, inflammation, and smooth muscle contraction.

14. Membrane structure is not a function of triacylglycerols. Membrane structure is a function of phospholipids.

16. HDL scavenges free cholesterol and transports it to the liver where it is converted to bile acids and excreted. This reduces total cholesterol in the serum and helps prevent plaque formation.

18. The sodium gradient created by the Na^+-K^+-ATPase in the plasma membrane of kidney tubule cells transports the glucose. This is an example of secondary active transport.

20. In passive transport, substances diffuse down a gradient. No energy is directly used to effect the transport. In active transport, energy is used to transport a substance against a concentration gradient. In simple diffusion, solutes move down a concentration gradient. Facilitated diffusion involves the movement of substances down a gradient via channels or carriers.

CHAPTER 11: ANSWERS TO EVEN-NUMBERED THOUGHT QUESTIONS

2. As time progresses, the antigens of the heterokaryon will intermingle. This suggests that the membrane is fluid and the antigens, as well as other components of the cell membrane, can move freely within the lipid bilayer.

4. The carbohydrate portion of the glycolipid can form hydrogen bonds with the water. This carbohydrate is the polar group, and is analogous to the charged portion of the phospholipid.

6. Cholesterol would stiffen the cell membrane. However, the presence of a cell wall would make incorporation of exogenous cholesterol into the underlying cell membrane improbable.

8. The ordered water molecules surrounding each phospholipid molecule are released from the polar heads. Order is lost and entropy increases.

CHAPTER TWELVE

Lipid Metabolism

FATTY ACIDS AND TRIACYLGLYCEROLS

The first step in the processing of dietary fat is the hydrolysis of triacylglycerols by pancreatic lipase and other intestinal lipases. This frees two fatty acids and leaves a monoacylglycerol.

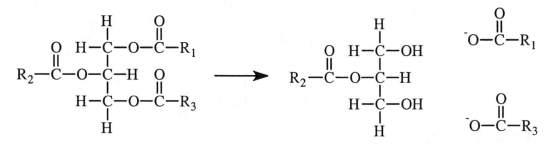

Free fatty acids have both polar and non-polar ends and can therefore function as biological "detergents" to solubilize triacylglycerols.

Further solubilization occurs in the intestine with bile salts. Bile salts are biological detergents made by the liver and secreted into the small intestine.

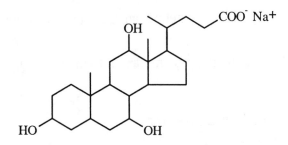

Sodium Cholate, a bile salt

Bile salts form mixed micelles with other fats. Micelles are then taken up into intestinal cells, reconverted to triacylglycerols, combined with cholesterol, phospholipids and protein to form chylomicrons, and transported out to the lymph, and then into the blood.

Glycerol must be transported back to the liver to be metabolized. In the liver, glycerol kinase converts it to glycerol-3-phosphate, which can be used in gluconeogenesis, or to make triacylglycerols or phospholipids.

Fatty acids can be used to synthesize triacylglycerols or membrane lipids, or degraded to generate energy.

135

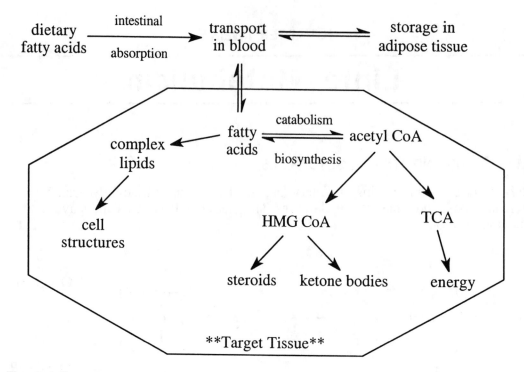

Target tissues:

The use of fatty acids for energy depends on the tissue. Nervous tissue, for instance, uses fatty acids minimally, but fatty acids are a major source of energy for muscle. During fasting, many tissues use fatty acids or ketone bodies for energy.

Fatty Acid Degradation (in the mitochondrial matrix)

1. Activation by acyl-CoA synthetase, an enzyme in the outer mitochondrial membrane

2. Transport (through the inner membrane into the matrix)

3. β-oxidation

ACTIVATION BY ACYL-COA SYNTHETASE

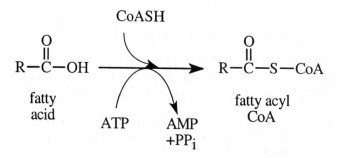

The ATP used to drive this reaction might be thought of as an "activation fee."

TRANSPORT: CARNITINE TRANSPORTS ACYL-CoA INTO THE MATRIX

Fatty acid catabolism occurs inside the mitochondrial matrix. Since fatty acyl CoA can't cross the inner membrane, the fatty acyl group is transferred to carnitine, which *can* be transported into the matrix.

The figure below shows the transport of palmitoyl-CoA through the inner membrane.[1] Carnitine and acyl-carnitine pass through the inner membrane with the help of a carrier protein. (Note that CoASH doesn't pass through the membrane, so the cell maintains separate cytoplasmic and mitochondrial pools of CoASH.)

Look closely at the acyl-carnitine structure. Can you see why it needs a carrier protein to get through the membrane?

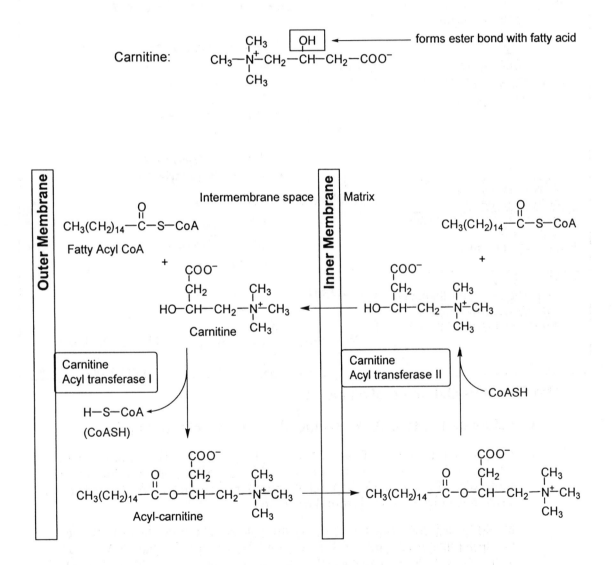

---

[1] Author's note: I thought that it would be worthwhile to see a real example. It's so convenient to use "R" for the long hydrocarbon chain that it can be easy to forget just how big these molecules are. *--PD*

β-OXIDATION

OF A FATTY ACYL-CoA WITH AN EVEN NUMBER OF CARBONS

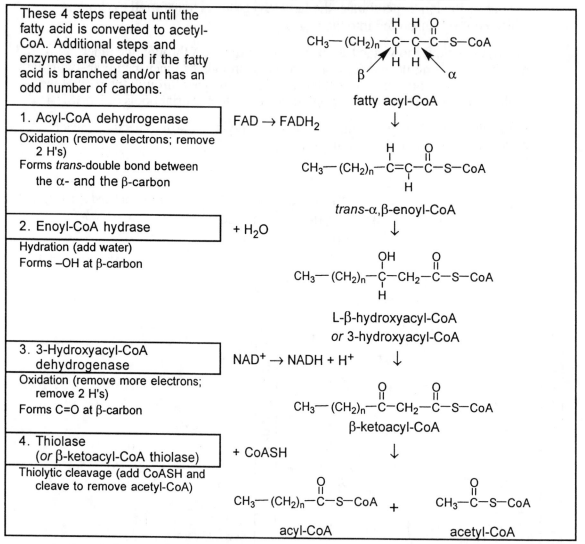

These 4 steps repeat until the fatty acid is converted to acetyl-CoA. Additional steps and enzymes are needed if the fatty acid is branched and/or has an odd number of carbons.

1. Acyl-CoA dehydrogenase

Oxidation (remove electrons; remove 2 H's)
Forms *trans*-double bond between the α- and the β-carbon

FAD → FADH₂

2. Enoyl-CoA hydrase

Hydration (add water)
Forms –OH at β-carbon

+ H₂O

3. 3-Hydroxyacyl-CoA dehydrogenase

Oxidation (remove more electrons; remove 2 H's)
Forms C=O at β-carbon

NAD⁺ → NADH + H⁺

4. Thiolase *(or β-ketoacyl-CoA thiolase)*

Thiolytic cleavage (add CoASH and cleave to remove acetyl-CoA)

+ CoASH

The Complete Oxidation of a Fatty Acid:

To calculate the total ATP produced per fatty acid molecule:

1. Determine the number of cycles of β-oxidation required. Each cycle produces one acetyl-CoA, one FADH₂, and one NADH; the last cycle produces two molecules of acetyl-CoA. List out the number of FADH₂, NADH, and acetyl-CoA molecules produced from β-oxidation.

2. Multiply each molecule above by the number of ATP it produces from the electron transport chain: 1.5 for each FADH₂ and 2.5 for each NADH. The acetyl-CoA molecule can enter the citric acid cycle to give 3 NADH, 1 FADH₂, and 1 GTP. So, each acetyl-CoA molecule produces 10 ATP (3x2.5 + 1.5 + 1).

3. From the total ATP produced, subtract 2 ATP to arrive at the final answer. (Those 2 ATP were the "activation fee" – the energy cost to activate the fatty acid. It counts as *two* ATP because an AMP, not an ADP, was formed.)

Palmitic acid (16 carbons) is described on page 382, and stearic acid (18 carbons) is Question 12.5. Try calculating how many NADH, FADH$_2$, and ATP molecules can be synthesized from 1 molecule of arachidic acid (20 carbons total).

Solution: 1.   Nine cycles of β-oxidation is required, producing 9 FADH$_2$, 9 NADH, and 10 acetyl-CoA.

2. $(9 \times 1.5) + (9 \times 2.5) + (10 \times 10) = 13.5 + 22.5 + 100 = 136$ ATP

3.   136 - 2 = 134 ATP total from one arachidic acid. Wow!

β-Oxidation in Peroxisomes *shortens very long-chain fatty acids*

Peroxisomal enzymes differ somewhat from the mitochondrial enzymes. Also, FADH$_2$ donates its electrons directly to O$_2$ to produce H$_2$O$_2$, which is converted to H$_2$O by catalase. The final enzyme doesn't bind medium-chain acyl-CoA's well; these are transported to the mitochondrion for further degradation.

Ketone Bodies: *ketogenesis occurs in the LIVER mitochondrial matrix*

KETOGENESIS: excess acetyl-CoA molecules are converted into ketone bodies.

acetoacetate   β-hydroxybutyrate   acetone

1. Acetoacetyl-CoA thiolase	acetyl-CoA acetyl-CoA
Condensation of two acetyl-CoA molecules Reverse of the last step of β-oxidation	↓↑ → CoASH acetoacetyl-CoA
2.HMG-CoA synthase	acetyl-CoA → ↓ → CoASH
Note that the acetyl-CoA added here will be regenerated in the next step, so this acetyl-CoA is really just a catalyst. Also requires H$_2$O.	HMG-CoA (β-hydroxy-β-methylglutaryl-CoA) ↓
3. HMG-CoA lyase	
(Compare this reaction to that of citrate synthase.)	acetyl-CoA + acetoacetate
4. β-hydroxybutyrate dehydrogenase	NADH + H$^+$ → NAD$^+$ ↓↑
Ketone bodies are used for energy by many tissues, most notably the heart and skeletal muscle. The brain prefers glucose for its fuel, but can use ketone bodies when it's forced to (i.e., during starvation or no-carbo diets). Ketone bodies are made in the liver, but the liver can't use them as fuel.	β-hydroxybutyrate

139

Ketosis is occurs when acetoacetate is made faster than it can be used. Under these conditions, acetoacetate spontaneously degrades to acetone (no enzyme is needed), which can be smelled as it's exhaled.	$CH_3-\overset{O}{\overset{\|}{C}}-CH_2-\overset{O}{\overset{\|}{C}}-O^- \rightarrow CH_3-\overset{O}{\overset{\|}{C}}-CH_3 + CO_2$

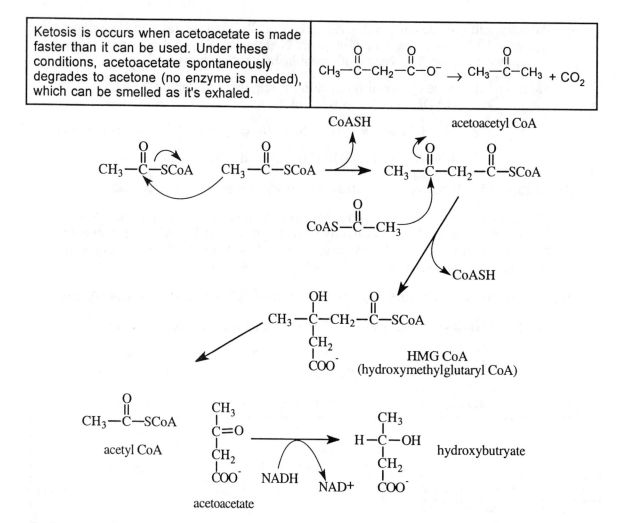

In the target tissues, acetoacetate is converted back to two molecules of acetyl-CoA:

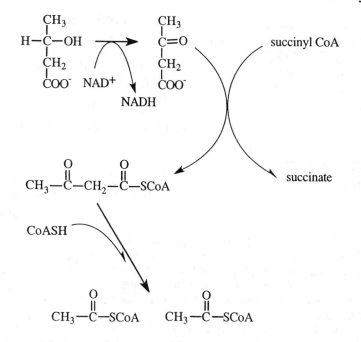

Fatty Acid Biosynthesis *takes place in the cytoplasm*

Fatty acid biosynthesis is not a simple reversal of β-oxidation because of thermodynamic considerations. (Both pathways must be spontaneous ($-\Delta G$) to go forward, but a simple reversal of any reaction will also reverse the sign of ΔG.)

Fatty acids are typically made from glucose: glucose → pyruvate → transported to mitochondrion → acetyl-CoA → citrate → citric acid cycle OR cross into the cytoplasm to make fatty acids.

1. Citrate transports acetyl-CoA from the mitochondrion to the cytoplasm.

2. Acetyl-CoA is activated by carboxylation to form malonyl-CoA (enzyme = acetyl-CoA carboxylase complex, which requires biotin, a CO_2 carrier).

3. Malonyl-CoA is transferred to an ACP = Acyl Carrier Protein. The malonyl-ACP serves as the carbon donor for fatty acid biosynthesis.

4. A second acetyl group is transferred from acetyl-CoA to synthase to form acetyl-S-synthase.

5. The synthase condenses the malonyl-ACP with the acetyl-S-synthase to form acetoacetyl-ACP; CO_2 is removed as well. This decarboxylation drives the process forward.

6. The reduction/dehydration/reduction gives the same intermediates as the β-oxidation of fatty acids.

7. The product butyryl-ACP transfers its acyl group to synthase, which condenses with another malonyl-ACP (again, with loss of CO_2). The process thus repeats until the fatty acid is synthesized (up to 16 carbons total).

CITRATE TRANSPORTS ACETYL-CoA FROM THE MITOCHONDRION TO THE CYTOPLASM

Acetyl-CoA is produced in the mitochondrion and must be transported to the cytoplasm, but CoASH can't cross the membrane. First, acetyl-CoA reacts with OAA to form citrate (remember that this is the first reaction of the citric acid cycle). Citrate crosses the mitochondrial membrane, then the reverse reaction occurs to free the acetyl-CoA.

Why would citrate leave the citric acid cycle in the mitochondrion and venture out into the cytoplasm? Remember that the citric acid cycle exists to produce energy. If energy isn't needed by the cell, citrate will accumulate.

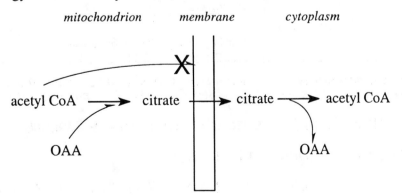

SUMMARY OF FATTY ACID BIOSYNTHESIS:

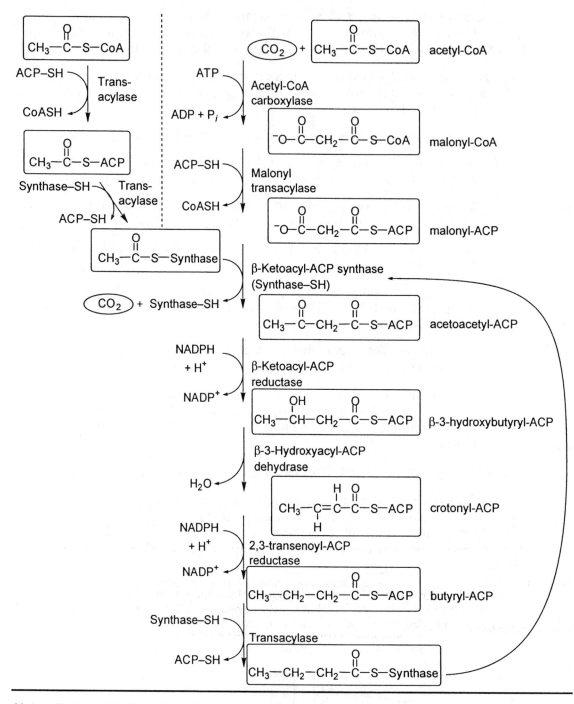

Note: Fatty acid fatty acid synthase is a multienzyme complex and a dimer, so two fatty acid molecules are synthesized at the same time. (See Figure 12.13 on page 393 of your text.)

THE OVERALL REACTION TO SYNTHESIZE A FATTY ACID WITH 16 CARBONS:

8 acetyl-CoA + 14 NADPH + 14 H$^+$ + 7 ATP $\rightarrow$

palmitate + 14 NADP$^+$ + 7 ADP + 7 P$_i$ + 8 CoASH + 6 H$_2$O

Comparison of fatty acid catabolism and biosynthesis

β-Oxidation

- occurs in the mitochondrion

- CoA is the acyl carrier

- FAD and NAD^+ are electron acceptors

- acetyl-CoA is the C-2 unit product

- enzymes differ

Biosynthesis

- occurs in the cytoplasm (chloroplasts in plants)

- ACP is the acyl carrier

- NADPH is the electron donor

- malonyl-CoA is the C-2 unit donor

- fatty acid synthase multienzyme complex - *dimer* (See Figure 12.13 on page 393 of your text.)

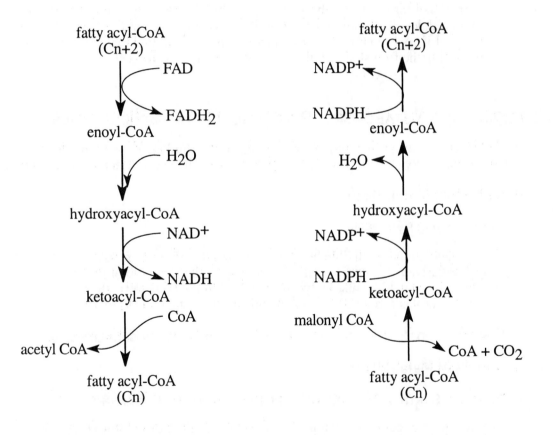

FATTY ACID ELONGATION AND DESATURATION - PRIMARILY BY ER ENZYMES

ER enzymes synthesize fatty acids longer than palmitate (16 carbons) and desaturates fatty acids (removes hydrogens to add double bonds) when these fatty acids aren't obtained by the diet.

Why is this important? Cells need unsaturated (and perhaps longer) fatty acids to maintain proper membrane fluidity, and as precursors for fatty acid derivatives (such as: eicosanoids; cerebrosides and sulfatides in myelin).

Regulation of Fatty Acid Metabolism in Mammals

HORMONE ACTION:

Glucagon, epinephrine: increases hydrolysis of triacylglycerols, so fatty acids enter the blood; inhibits first step of fatty acid synthesis

Insulin: promotes lipogenesis, inhibits lipolysis

SUBSTRATES AND PRODUCTS:

Citrate stimulates the first step of fatty acid synthesis (acetyl-CoA carboxylase).

Palmitoyl-CoA inhibits fatty acid synthesis and the pentose phosphate pathway.

Malonyl-CoA inhibits the transport of fatty acids into the mitochondrion for β-oxidation (i.e., it inhibits carnitine acyltransferase I). Why? Remember that the first step of fatty acid synthesis makes malonyl-CoA. When that first step is inhibited, levels of malonyl-CoA drop. Since it's not there to act as inhibitor, the acyl-CoA can enter the mitochondrion to be degraded by β-oxidation. So, inhibiting the first step of fatty acid synthesis propels β-oxidation.

MEMBRANE LIPID METABOLISM: PHOSPHOLIPIDS AND SPHINGOLIPIDS

REMODELING: a process that allows a cell to adjust the fluidity of its membranes. Unsaturated fatty acids replace the original fatty acids incorporated during synthesis.

Phospholipid Metabolism

PHOSPHOLIPID SYNTHESIS

Once ethanolamine or choline has entered a cell, it is phosphorylated and converted to a CDP derivative. Subsequently, phosphatidylethanolamine or phosphatidylcholine is formed when diacylglycerol reacts with the CDP derivative. Triacylglycerol is produced when diacylglycerol reacts with acyl-CoA.

PHOSPHOLIPID DEGRADATION IS CATALYZED BY SEVERAL PHOSPHOLIPASES.

Sphingolipid Metabolism

SPHINGOLIPID SYNTHESIS BEGINS WITH THE PRODUCTION OF CERAMIDE:

The synthesis of ceramide begins with the condensation of palmitoyl-CoA with serine to form 3-ketosphinganine. 3-Ketosphinganine is subsequently reduced by NADPH to form sphinganine. In a two-step process involving acyl-CoA and $FADH_2$, sphinganine is converted to ceramide. Ceramide can then be used to form sphingomyelin, glucosylceramide, or galactosylceramide.

SPHINGOLIPID DEGRADATION OCCURS IN LYSOSOMES BY SPECIFIC HYDROLYTIC ENZYMES.

ISOPRENOID METABOLISM

Isoprenoids occur in all eukaryotes. Despite the astonishing diversity of isoprenoid molecules produced, there's a great deal of similarity in the synthesizing mechanism of different species. In fact, the initial phase of isoprenoid synthesis (the synthesis of isopentyl pyrophosphate) appears to be identical in all of the species in which this process has been investigated.

Cholesterol Metabolism

The synthesis of cholesterol can be divided into three phases:

   a. formation of HMG-CoA (β-hydroxy-β-methylglutaryl-CoA) from acetyl-CoA,

   b. conversion of HMG-CoA to squalene, and

   c. conversion of squalene to cholesterol.

The most important mechanism for degrading and eliminating cholesterol is the synthesis of bile acids. The conversion of cholesterol to 7-α-hydrocholesterol, catalyzed by cholesterol-7-hydroxylase, is the rate-limiting reaction in bile acid synthesis. Subsequent reactions result in the rearrangement and reduction of double bond at C-5, the introduction of an additional hydroxyl group, and the reduction of the C-3-keto group to a 3-α-hydroxyl group. The products of this process, cholic acid and deoxycholic acid, are converted to bile salts by microsomal enzymes which catalyze conjunction reactions.

The amount and types of steroids synthesized in a specific tissue are carefully regulated. Cells in each tissue are programmed during embryonic and fetal development to respond to a variety of chemical signals by inducing the synthesis of a unique set of specific enzymes. The most important chemical signals that are now believed to influence steroid metabolism are various peptide hormones secreted from the pituitary and several prostaglandins.

CHAPTER 12: ANSWERS TO EVEN-NUMBERED REVIEW QUESTIONS

2.   a. Carnitine is an amino acid that is used to transport acyl-CoA molecules into the mitochondria.

   b. Flippase is a protein that transfers choline-containing phospholipids across the membrane.

   c. Thrombin is a proteolytic enzyme that converts fibrinogen to fibrin.

   d. Thiolase catalyzes the final reaction in the β-oxidation cycle, referred to as a thiolytic cleavage, which yields an acetyl-CoA product.

   e. Desmolase catalyzes the initial reaction in steroid hormone synthesis, the conversion of cholesterol to pregnenolone.

   f. Phospholipid exchange protein is a water-soluble protein that binds phospholipids in one membrane and transfers them to another membrane.

   g. Sterol carrier protein is a cytoplasmic molecule that binds squalene during cholesterol synthesis.

   h. ACTH, adrenocorticotropic hormone, is a peptide hormone released by the pituitary gland that stimulates the synthesis of adrenal steroids.

   i. Glucocorticoids are hormones that promote carbohydrate, protein, and fat metabolism.

4. Three differences between fatty acid synthesis and β-oxidation are the following: (1) The two pathways take place in different cell compartments. Synthesis occurs in the cytoplasm and β-oxidation is within the mitochondria. (2) The intermediates of fatty acid synthesis and β-oxidation are linked through thioester linkages to ACP and CoASH respectively. (3) The electron carrier for fatty acid synthesis is NADPH while those for β-oxidation are NADH and $FADH_2$.

6. In the short term, hormones alter the activity of preexisting regulatory enzyme molecules. For example, the binding of glucagon inhibits acetyl-CoA carboxylase. Long-term effects of hormones usually involve changes in the pattern of enzyme synthesis in target cells. For example, insulin promotes the synthesis of the enzymes involved in lipogenesis (e.g., acetyl-CoA carboxylase and fatty acid synthase).

8. Because of the presence of a methyl substituent on the β-carbon, the fatty acid first undergoes one cycle of α-oxidation. The resulting molecule, now shorter by one carbon atom, then undergoes one cycle of β-oxidation. The products of this latter process are two molecules of propionyl-CoA.

10. Enoyl-CoA isomerase converts the naturally occurring *cis* double bond at Δ^3 to a *trans* double bond at Δ^2, the correct position for the next round of β-oxidation.

12. The indicated bond is cleaved by glucocerebrosidase.

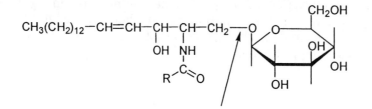

14. Review Figure 12.28 on p 412. The bile salts are emulsifying agents that facilitate fat digestion.

CHAPTER 12: ANSWERS TO EVEN-NUMBERED THOUGHT QUESTIONS

2. The potential consequences of faulty regulation include the creation of some level of futile cycling in which energy is wasted and the requirements of the cell for fatty acid synthesis and energy generation are compromised.

4. a. Hydrophobic interactions are probable between the enzyme and the lipid in the micelle.

   b. For phospholipases to be drawn into the micelle, they must have a hydrophobic surface.

6. Although regular eating is not a panacea, it does provide sufficient carbohydrate to act as a fuel to sustain vital metabolic processes.

8. In periods of fasting, blood glucose levels fall and glucagon and epinephrine are released. These hormones then bind to their respective adipocyte plasma membrane receptors. This binding initiates a cascade (cAMP activates protein kinase which in turn activates hormone sensitive lipase) that results in the release of fatty acids and glycerol into the blood.

CHAPTER THIRTEEN

Photosynthesis

PHOTOSYNTHESIS: WHAT ARE THE ULTIMATE GOALS?

- to convert light energy into chemical energy (ATP)

- to produce reducing equivalents (NADPH)

- to convert CO_2 into sugars

LIGHT ENERGY drives the production of **ATP** and **NADPH**. The ATP and NADPH produced are used to convert CO_2 into **SUGARS**.

STUDY HINTS

To make sense out of photosynthesis, we need to understand and integrate all of the following topics, and each topic contains new terms and new concepts. The good news is that you're already familiar with a fair amount of this, and there are a number of similarities between systems that you've studied before (especially the electron transport system) and those involved in photosynthesis.

- Chloroplast structure
- How light excites electrons, and what those excited electrons can do
- Molecular structures and complexes that are receptive to light (chromophores)
- Electron transfer and the use of the energy released as a result
- ATP as energy currency in the cell; NADPH as reducing power
- Metabolism of carbohydrate molecules and regulation of pathways

STAY FOCUSED on the goal of each part of photosynthesis that you're studying, and how it relates to the ultimate goals listed above. A stumbling block to avoid is to become so mired in the details of one section that you lose sight of its importance. Understand the big picture first, then tackle the specifics.

Example: The *Z Scheme* is a new term, but it's just another way of looking at photosynthesis. Compare Figure 13.10, which outlines the Z scheme, to Figure 13.14, which is a general schematic of the light reactions of photosynthesis.

CHLOROPLAST STRUCTURE: Knowing the location (i.e., stroma, thylakoid lumen, membrane) and how the parts of the chloroplast fit together will really help to give a clear picture of the overall process, and will also help to clarify how the different photosystems and the dark reactions fit together. Keep a diagram of the chloroplast in front of you, and when terms such as "thylakoid membrane" and "stroma" pop up, you'll have an immediate visual. Figure 13.2 on page 421 is great. Be sure to include *appressed* and *nonappressed* in your diagram.

Example: Since the ATP synthase complex works by allowing H^+ to pass through its pore (and as such, is driven by an $[H^+]$ gradient, just like mitochondrial ATP synthase), it's significant *where* each H^+ is used or released by a particular reaction.

ABBREVIATIONS: Many of the molecules, complexes, and systems have long names, and most of them have abbreviations. To stay afloat in this alphabet soup, keep a running list of abbreviations that you don't know right off the top of your head, and keep this list in front of you as well. *(Examples include PSII, PQ, Y_z, MSP, and LHCII.)*

Additional Note Regarding Abbreviations: Some molecules have more than one abbreviation. It's valuable to be aware of all of the abbreviations (since different sources/texts have different preferences), but don't let them frustrate the learning process. Have a good concise list on hand. Example: Ferredoxin-NADP oxidoreductase is referred to as both FP and FNR, and pheophytin a is shown as both Ph and Phe a.

A PICTURE IS WORTH A THOUSAND WORDS: Make your own diagrams, using the figures in the text as a guide. The drawing and writing process is a powerful learning tool. These summary figures are particularly good:

Figure 13.14   *Membrane Organization of the Light Reactions in Chloroplasts: the Electron Transport Chain and the ATP Synthase Complex.* This figure shows the interactions of light with photosystems I and II, their connection, and the roles of the cyt b_6F complex, PQ, and ATP synthase, on location in and around the thylakoid membrane.

Figure 13.2   *Chloroplast structure.* Again, be sure to differentiate between appressed and nonappressed thylakoid membrane in your diagrams.

Figures 13.10-11   *The Z Scheme* and *More Details of the Z Scheme.*

Figure 13.15   *The Calvin Cycle:* using CO_2, ATP, and NADH to make a 3-carbon sugar. What is the net equation, and what happens at each phase? Note that parts of this cycle resembles the pentose phosphate pathway.

Figure 13.16   *Photorespiration.*

CHLOROPLAST STRUCTURE *(HINT: COMPARE CHLOROPLASTS TO MITOCHONDRIA)*

STROMA - space inside the inner membrane (similar to mitochondrial matrix)

contains enzymes (for light-independent reactions and starch synthesis), DNA, ribosomes

THYLAKOID MEMBRANE - third membrane - folded into series of grana

within thylakoid membrane: light-dependent reactions of photosynthesis

APPRESSED THYLAKOID MEMBRANE - adjacent layers of membrane that fit closely together within each granum

NONAPPRESSED THYLAKOID MEMBRANE is directly exposed to the stroma.

GRANUM - stack of several flattened vesicles (plural = grana)

THYLAKOID LUMEN (SPACE) - internal compartment created by formation of grana

STROMAL LAMELLA - thylakoid membrane that interconnects grana

LIGHT HAS PROPERTIES OF BOTH WAVES AND PARTICLES

VISIBLE LIGHT:

violet		*red*
400 nm	to	700 nm
shorter wavelength		longer wavelength
higher frequency		lower frequency
higher energy		*lower energy*

WAVE PROPERTIES OF LIGHT: LIGHT ACTS LIKE WAVES	PARTICLE PROPERTIES OF LIGHT: PHOTONS LIGHT ACTS LIKE PARTICLES
$\lambda = c/v$ λ = wavelength, v = frequency, c = speed of light	$\varepsilon = hv$ ε = energy of a photon h = Planck's constant

Molecules absorb light at specific energies that correspond to specific wavelengths. (Complex molecules can absorb light at several wavelengths.)

Chromophores absorb light. When electrons in the chromophores absorb light, they become excited (move from the ground state to higher energy levels).

HOW AN EXCITED ELECTRON CAN RETURN TO ITS GROUND STATE:

1. **FLUORESCENCE** - electron relaxes to a lower vibrational (energy) state, then goes back to ground state and emits a photon (of lower energy than original photon absorbed)

2.* **RESONANCE ENERGY TRANSFER** - excitation energy is transferred to a neighboring molecule via resonance

3.* **OXIDATION-REDUCTION** - transfer of an excited electron to a neighboring molecule, making it a strong reducing agent. The electron returns to its ground state by reducing another molecule.

4. **RADIATIONLESS DECAY** - excitation energy is given off as heat

*These two methods are the most important to photosynthesis.

Light energy is converted into chemical energy at reaction centers. Each reaction center is a complex of light-absorbing pigments and electron transfer proteins. Chlorophyll a, chlorophyll b, and carotenoids are specialized pigment molecules that absorb light energy. (Chromophores that absorb visible light typically have extended chains of conjugated alkenes.)

LIGHT HARVESTING COMPLEX II (LHCII) - transmembrane protein that binds many chlorophyll a, chlorophyll b, and carotenoid molecules; major component of thylakoid membrane

ANTENNA PIGMENTS - molecules that absorb light energy and transfer it to the reaction center

LIGHT REACTIONS: LIGHT EXCITES ELECTRONS THAT ARE USED TO SYNTHESIZE ATP AND NADPH

OVERALL REACTION: H_2O IS OXIDIZED TO O_2, $NADP^+$ IS REDUCED TO NADPH

$$2\,NADP^+ + 2\,H_2O \Leftrightarrow 2\,NADPH + O_2 + 2\,H^+ \qquad \Delta E_0' = -1.136\ V$$

For each mole of O_2 generated:

at least 438 kJ of free energy ($\Delta G^{\circ\prime}$) is needed

at least 8 photons are absorbed to provide 1360 kJ of energy

SUMMARY OF LIGHT REACTIONS

Light is used in photosystem II to excite electrons to a higher energy state. As the electrons are transferred down an energy gradient, protons (H^+) are pumped across the membrane to generate a proton gradient, which is used for ATP synthesis.

The electrons from photosystem II can then be transferred to photosystem I. More light is needed to excite the electrons to an energy state that's high enough to reduce $NADP^+$ to NADPH.

Water serves as a source of electrons when electrons are transferred from photosystem II to photosystem I to $NADP^+$. When H_2O gives up its electrons, the oxygen in H_2O is oxidized to O_2.

Z SCHEME: ELECTRON TRANSFER FROM H_2O TO $NADP^+$

The Z scheme shows where electrons go during photosynthesis. If this path of electrons is shown on a graph where the y axis is energy, then part of this path resembles the letter Z.

Light drives photosynthesis, and the Z scheme shows where and why this is true. From the energy differences between molecules within the photosystems, it's clear that a boost of energy - in the form of a photon of light - is needed at PSII and at PSI to give the electrons enough energy to continue through their path to $NADP^+$.

Overall, electrons come from H_2O and reduce $NADP^+$ to form NADPH.

PATH OF ELECTRONS THROUGH THE PHOTOSYSTEMS IN THE PRESENCE OF LIGHT:

$$H_2O \rightarrow \boxed{PSII} \rightarrow PQ \rightarrow cyt\ b_6f \rightarrow PC \rightarrow \boxed{PSI} \rightarrow Fd \rightarrow FNR \rightarrow NADP^+$$

$H_2O \rightarrow$ photosystem II $\rightarrow$ plastoquinone $\rightarrow$ cytochrome b_6f complex $\rightarrow$ plastocyanin $\rightarrow$ photosystem I $\rightarrow$ ferredoxin $\rightarrow$ ferredoxin-NADP oxidoreductase $\rightarrow$ $NADP^+$

Another way to think about this is to begin with the excitation of photosystem I, which results in the transfer of electrons, ultimately to $NADP^+$ to form NADPH. After the electrons are released from PSI, how are they replaced?

When P700 in PSI absorbs a photon of light, it releases an energized electron. That electron is replaced by an electron from PSII (through PQ, cyt b_6f, and PC, as shown above). The PSII electron needs to be excited by a photon as well, in order to have enough energy to be transferred. How is the PSII electron replaced? *Water gives up electrons to form O_2 and H^+.*

Photosystem II and Oxygen Generation

PS II is a large membrane-spanning protein-pigment complex with at least 23 components.

The function of photosystem II is to oxidize water molecules and donate energized electrons to electron carriers which eventually reduce photosystem I. As the electrons are donated to the electron carriers, protons are pumped out of the thylakoid and ultimately used for ATP synthesis. The water-oxidizing clock is the mechanism by which H_2O is converted into O_2.

Location: primarily in thylakoid membrane with grana (appressed membrane not exposed to stroma)

Reactions of photosystem II:

Light energy excites a P680 electron, giving it a large potential energy. These energized electrons are transferred to a series of electron transport carriers that are analogous to the mitochondrial electron transport chain. As the electrons are transferred, protons are pumped from the stroma into the thylakoid lumen. This creates a proton gradient that results in the synthesis of ATP.

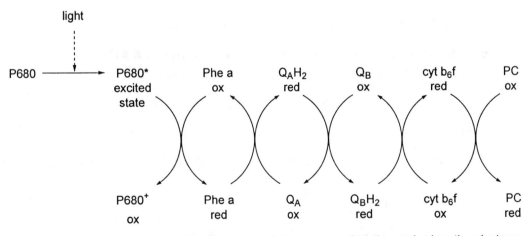

Note: The reduced state means that the carrier has the electrons.

THE REDUCED PLASTOCYANIN TRANSFERS ITS ELECTRONS TO P700 IN PS I.

WHAT HAPPENS TO THE OXIDIZED P680?

These electrons are replaced when H_2O is oxidized.

PQ transfers electrons from PS II to cyt b$_6$f and pumps H$^+$ to the lumen

PQ is plastoquinone, an electron carrier within the thylakoid membrane

$$PQ + 2\ e^- \text{ (from PS II)} + 2\ H^+ \text{ (from stroma)} \rightarrow PQH_2$$

$$PQH_2 \rightarrow PQ + 2\ e^- \text{ (to cyt b}_6\text{f)} + 2\ H^+ \text{ (to the thylakoid lumen)}$$

Cytochrome b$_6$f complex transfers electrons from PQ to PC

Location: throughout the thylakoid membrane

An Fe-S site on the complex transfers electrons from PQ to PC. PC is plastocyanin, a water-soluble, copper-containing protein in the thylakoid lumen.

Photosystem I uses electrons from PC to reduce NADP$^+$ → NADPH

The ultimate goal of PS I is to reduce NADP$^+$ to NADPH, which requires two electrons.

PSI begins with the transfer of an electron from PC (in the lumen) to P700 (in the membrane).

This reduced P700 then absorbs a photon of light that excites and energizes this electron.

The excited electron is then transferred through this series of electron carriers:

$P700^* \rightarrow A_0 \rightarrow Q \rightarrow F_x \rightarrow F_A, F_B$ (These are all in PS I, in the thylakoid membrane.)

chlorophyll a → phylloquinone → Fe-S proteins

The next electron acceptor is ferredoxin, in the stroma.

Ferredoxin transfers its electron to FNR (to make NADPH) or to PQ (to pump H$^+$ for ATP synthesis)

Ferredoxin is mobile, water-soluble, and in the stroma.

FERREDOXIN → FNR IS THE *NONCYCLIC* ELECTRON TRANSPORT PATHWAY

Ferredoxin transfers energized electrons from PSI to FNR (ferredoxin-NADP oxidoreductase).

FNR uses two electrons (and a stromal H$^+$) to reduce NADP$^+$ to NADPH.

FERREDOXIN → PQ IS THE *CYCLIC* ELECTRON TRANSPORT PATHWAY

Electrons are cycled *back* to PQ (in the thylakoid membrane), which also takes two H$^+$ from the stroma to form PQH$_2$. Remember that PQH$_2$ then gives its electrons to cyt b$_6$f and its two H$^+$ go to the lumen.

So, energized electrons were used to pump additional H$^+$ across the thylakoid membrane into the lumen *instead of making NADPH*.

The extra H$^+$ in the lumen produces additional ATP.

The cyclic electron transport pathway drives ATP synthesis without making NADPH. This only occurs in some species (e.g., algae), when the $NADPH/NADP^+$ ratio is high.

ATP synthase = CF_0CF_1ATP synthase - phosphorylates ADP, driven by transmembrane proton gradient (produced during light-driven electron transport)

Location: thylakoid membrane that's directly in contact with the stroma

Photophosphorylation is the light-driven synthesis of ATP from ADP + P_i.

LIGHT-INDEPENDENT REACTIONS

The Calvin Cycle[1] incorporates CO_2 into carbohydrate molecules

Many Calvin cycle reactions resemble pentose phosphate pathway reactions.

PHASES OF THE CALVIN CYCLE:

1. **CARBON FIXATION** by the enzyme rubisco (ribulose-1,5-bisphosphate carboxylase) in C3 plants: CO_2 reacts with ribulose-1,5-bisphosphate to form 2 glycerate-3-phosphates (per CO_2). This reaction requires ATP.

2. **REDUCTION OF GLYCERATE-3-PHOSPHATE** by NADPH to form glyceraldehyde-3-phosphate

3. **REGENERATION OF RIBULOSE-1,5-BISPHOSPHATE**: Of every six glyceraldehyde-3-phosphates formed, five are regenerated to form three molecules of ribulose-1,5-bisphosphate.

CALVIN CYCLE REACTION SUMMARY:

3 ribulose-1,5-bisphosphate + 3 CO_2	$\rightarrow$	6 glycerate-3-phosphate
6 glycerate-3-phosphate + 6 ATP + 6 NADPH	$\rightarrow$	6 glyceraldehyde-3-phosphate + 6 ADP + 6 $NADP^+$ + 6 P_i
5 glyceraldehyde-3-phosphate + 3 ATP	$\rightarrow$	3 ribulose-1,5-bisphosphate + 3 ADP + 2 P_i

CALVIN CYCLE NET EQUATION:

3 CO_2 + 6 NADPH + 9 ATP	$\rightarrow$	glyceraldehyde-3-phosphate + 6 NADP+ + 9 ADP + 8 P_i

---

[1] The Calvin cycle is also known as: the dark reactions, light-independent reactions, the reductive pentose phosphate cycle (RPP cycle) and the photosynthetic carbon reduction cycle (PCR cycle.)

Photorespiration consumes O_2 and liberates CO_2

Photorespiration is a wasteful light-dependent process that undermines photosynthesis. In photorespiration, O_2 is consumed and CO_2 is evolved by plant cells which are actively engaged in photosynthesis.

The rate of photorespiration depends on the concentrations of CO_2 and O_2, and its function is unknown. Plants use C4 metabolism and crassulacean acid metabolism (CAM) to counteract the photorespiration process. (See further notes, below.)

REGULATION OF PHOTOSYNTHESIS: RUBISCO CONTROL; EFFECT OF LIGHT

Goals: To increase the rate of photosynthesis when light is available, and to make sure that when the rate of photosynthesis is high, the rates of CO_2 fixation and sucrose synthesis is also high.

Photosynthesis depends on temperature, cellular CO_2 concentration and light. Light activates certain photosynthetic enzymes and deactivates some of the enzymes that degrade sugars.

The key regulatory enzyme in photosynthesis is ribulose-1,5-bisphosphate carboxylase (rubisco). The reaction that rubisco catalyzes, the fixation of CO_2, is relatively slow. The best way to increase its rate is to increase the number of copies of the enzyme. Genes that code for rubisco are activated by an increase in light intensity (and appears to involve phytochrome).

Rubisco is also controlled by covalent modification (carbamoylation). The rate of carbamoylation depends on the CO_2 concentration and an alkaline pH.

Light Affects Enzymes by Indirect Mechanisms:

1. **pH** of the stroma increases during photosynthesis (when H^+ is pumped out of the stroma into the thylakoid lumen). This higher pH (~ 8) activates rubisco and other enzymes.

2. **Mg^{2+}** moves across the membrane into the stroma during the light reactions; This increase in stromal $[Mg^{2+}]$ activates several photosynthetic enzymes (including rubisco).

3. FERREDOXIN - THIOREDOXIN SYSTEM

4. PHYTOCHROME

ALTERNATIVES TO C3 METABOLISM: C4 METABOLISM AND CAM (CRASSULACEAN ACID METABOLISM)

C4 plants have evolved mechanisms to reduce water loss, and photorespiration. C4 metabolism is a mechanism for assimilating CO_2.

When the plant opens its stomata at night, CO_2 enters and reacts to form oxaloacetate. When light is available, photosynthesis makes ATP and NADPH, and CO_2 is released in the bundle sheath cells and converted into sugar. Because the concentration of CO_2 within bundle sheath cells is significantly higher than that of O_2, photorespiration is drastically reduced.

In C4 plants, the light reactions occur in the mesophyll cells, and the Calvin cycle reactions (CO_2 fixation) occur in the bundle sheath cells. This division of labor lets C4 plants concentrate CO_2 in the bundle sheath cells at the expense of ATP hydrolysis.

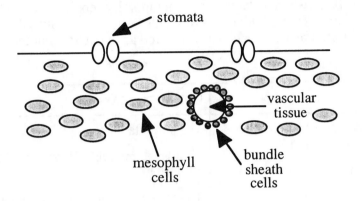

ADVANTAGES OF C4 METABOLISM

- Minimizes photorespiration by increasing CO_2 concentration in cells that have rubisco.

- Reduces the plant's need for water. Stomata must open to allow CO_2 to enter and O_2 to exit the leaf. But water is also lost when the stomata are open. C4 plants can effectively concentrate CO_2, so that the stomata do not need to be open as often. Water loss in C4 plants is only 10-30% compared to C3 plants.

MESOPHYLL CELLS[2]	BUNDLE SHEATH CELLS
IN DIRECT CONTACT WITH AIR	IN CONTACT WITH VASCULAR TISSUE
LACK RUBISCO	HAVE RUBISCO
LIGHT REACTIONS MAKE **ATP, NADPH**	CALVIN CYCLE REACTIONS USE CO_2 RELEASED

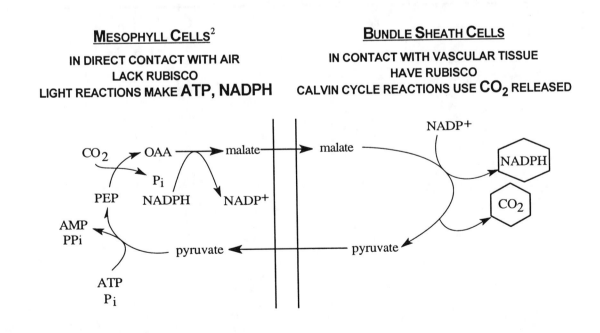

---

[2] In some C4 plants, oxaloacetate is converted to aspartate instead of malate. The Asp travels to the bundle sheath cells, where it's converted back to oxaloacetate, which can then release its CO_2.

CAM (Crassulacean Acid Metabolism)

CAM is another type of photosynthetic specialization used by desert plants. These plants grow in high intensity light without very much water and, in order to survive, have evolved to the point where they are able to temporally separate carbon fixation and ATP/NADPH synthesis. The growth of CAM plants is slowed because the extent of photosynthesis is limited by how much CO_2 is stored as malate during the night.

AT NIGHT:	DURING THE DAY:
The stomata open at night when water loss would be at a minimum. CO_2 is stored as malate.	The stomata close. Photosystems I and II work to make ATP and NADPH. The CO_2 that was stored in malate during the night is released, and rubisco uses this CO_2 to make carbohydrates in the Calvin Cycle.

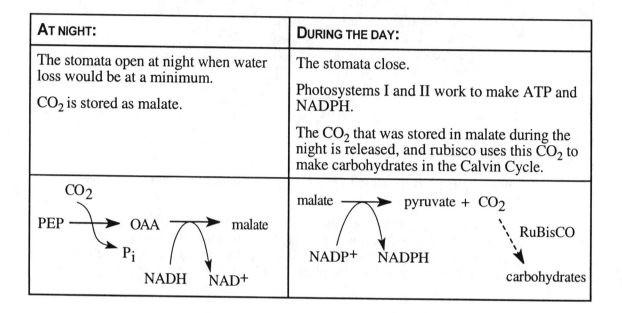

CHAPTER 13: ANSWERS TO EVEN-NUMBERED REVIEW QUESTIONS

2. The most significant contribution of early photosynthesizing organisms to the earth's environment was the conversion of a reducing atmosphere (ammonia and methane) to an oxidizing atmosphere.

4. Chloroplasts resemble mitochondria in the following ways: (1) they are both similar in size and structure to modern prokaryotes; (2) they both reproduce by binary fission; (3) the genetic information and protein synthesizing capability of both chloroplasts and mitochondria are similar to that of prokaryotes; (4) the ribosomes of chloroplasts and mitochondria are similar in size and function; and (5) they are both thought to have arisen from ancient free living prokaryotes.

6. The final electron acceptor in photosynthesis is carbon dioxide.

8. The net production of the dark, or light-independent, reactions of photosynthesis is one molecule of glyceraldehyde-3-phosphate. See page 435 for the reactions of the Calvin cycle.

10. The maximum rate of photosynthesis will occur at the λ_{max} of the photosynthesizing system. This absorption maximum should match the absorption maxima of the light absorbing pigments. This topic is described in Biochemical Methods 13.1.

12. If blue wavelengths are used in addition to red ones, the rate of oxygen evolution is increased. This phenomenon is known as the Emerson enhancement effect. If photosynthesis occurs in a single photosystem, then the magnitude of the enhancement should reflect the ratios of the λ_{max} in the red and blue regions. The enhancement was much greater than predicted by this ratio. The existence of a second chromophore (and by inference a second photosystem) was therefore suggested.

14. The oxygen-evolving complex of PSII exists in five transient oxidation states (S_0 through S_4), collectively referred to as a clock. Because oxygen evolution occurs only when the S_4 state has been reached, several light bursts are required.

16. Carbon dioxide fixation occurs in the stroma of the chloroplasts.

CHAPTER 13: ANSWERS TO EVEN-NUMBERED THOUGHT QUESTIONS

2. Conjugation is a system of alternate double and single bonds. When light with sufficient energy strikes the π electrons of a conjugated system (or any double bond), an electron is promoted from the ground state to a higher energy state, referred to as the excited state. Conjugation lowers the energy difference between the ground state and the excited state, hence photons of lower energy are capable of achieving this transition.

4. Oxidative phosphorylation and photophosphorylation use many of the same molecules in their reactions and both are linked to an electron transport system. However, chloroplasts use light energy to drive redox reactions while mitochondria use the energy of chemical bonds to drive redox reactions. In contrast to mitochondrial inner membrane, thylakoid inner membrane is permeable to

magnesium and chloride ions. Therefore, the electrochemical gradient across the thylakoid membrane consists mainly of a proton gradient.

6.   If sufficient carbon dioxide is already present to saturate all of the ribulose- 1,5-bisphosphate carboxylase molecules, the presence of additional carbon dioxide molecules will not increase the rate of photosynthesis. In addition, photosynthesis is depressed by low light levels.

8.   Photorespiration is carried out by RuBisCO, an enzyme that has both oxygenase and carbon dioxide fixation activities. Under conditions of high carbon dioxide concentration, the oxygenase functions are repressed and photorespiration slows down.

10.   Because of the capacity of C4 plants to avoid the process of photorespiration, herbicides that promote photorespiration do not affect these organisms.

Nitrogen Metabolism I: Synthesis

NITROGEN FIXATION ($N_2 \rightarrow NH_3$) IS A REDUCTION THAT REQUIRES ENERGY

Requires energy: at least 16 ATP per N_2

Nitrogenous Complex includes: *(must be anaerobic – protected from O_2)*

Dinitrogenase (or Fe-Mo protein): $N_2 + 8\,H^+ + 8\,e^- \rightarrow 2\,NH_3 + H_2$

Dinitrogenase reductase (or Fe protein): binds ATP; ATP hydrolyzes to ADP and P_i; this causes conformational changes that help the e^- transfer to dinitrogenase

Path of electrons:

NADH (or NADPH) $\rightarrow$ ferredoxin $\rightarrow$ dinitrogenase reductase $\rightarrow$ dinitrogenase

NH_3 then travels from the bacteria to the host cells that use it to synthesize glutamine.

Nitrogen fixation allows many plants and animals to synthesize many N-containing biomolecules such as proteins and nucleic acids. Only a few prokaryotes can fix nitrogen. Plants receive their nitrogen via symbiotic relationships with N_2-fixing prokaryotes or by absorbing NH_3 and NO_3^- synthesized by soil bacteria (or provided by artificial fertilizers). Animals take in organic nitrogen mainly as amino acids. The liver determines the fate of ingested amino acids.

OUR SOURCE OF NITROGEN: AMINO ACIDS

BCAA Branched chain amino acids   Leu, Ile, Val

EAA Essential amino acids must be provided by the diet.   Ile, Leu, Lys, Met, Phe, Thr, Trp, Val

NAA Nonessential amino acids can be synthesized from available metabolites.   Ala, Arg*, Asn, Asp, Cys, Glu, Gln, Gly, His*, Pro, Ser, Tyr

*essential for infants

Hint:

Now would be a great time to brush up on the amino acid R groups! It'll be a tremendous help to you if their structures instantly pop into your mind as you study.

AMINO ACID BIOSYNTHESIS

Amino Acid Metabolism Overview

Amino acid pool = amino acids available for metabolic processes. Amino acids from the degradation of dietary and tissue proteins enter the pool; excreted nitrogenous end products such as urea and uric acid leave the pool. Amino acids enter cells via membrane-bound transport proteins; some are Na^+-transport-dependent.

Nitrogen balance: nitrogen intake = nitrogen loss

Positive nitrogen balance: nitrogen intake > nitrogen loss; Protein synthesis exceeds degradation (growing children, pregnant women, recuperating patients)

Negative nitrogen balance: nitrogen intake < nitrogen loss; can't replenish nitrogen fast enough; malnutrition (Kwashiorkor)

Reactions of Amino Groups

TRANSAMINATION REACTIONS: THE TRANSFER OF AMINO GROUPS MAKES THE SYNTHESIS OF NEW AMINO ACIDS POSSIBLE

Amino groups are transferred **from an α-amino acid to an α-keto acid.** (Remember that "α-acid" places these groups *right next to* a carboxylic acid.) Because transamination reactions are readily reversible, they play an important role in both the synthesis and degradation of the amino acids.

Transamination Example: Remember the alanine cycle?

pyruvate   glutamate   alanine α-ketoglutarate

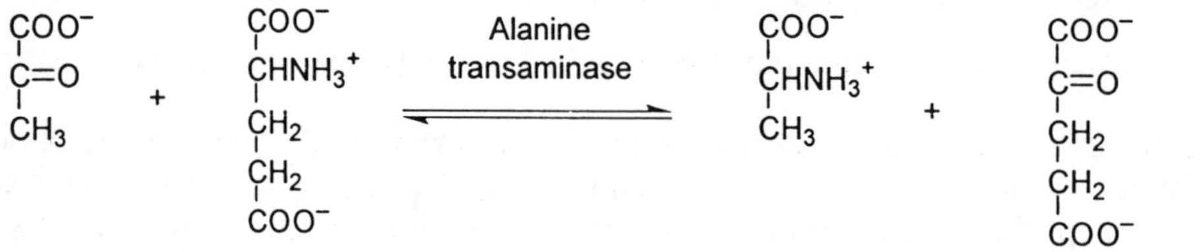

AMINOTRANSFERASES:

Aminotransferases require the coenzyme pyridoxal-5′-phosphate (PLP) (from pyridoxine / vitamin B_6) that forms a Schiff base ($-C=N-$) between its aldehyde C and the N of the amino acid's $-NH_3^+$.

Two types of aminotransferases:

1. Specific for the type of α-amino acid that donates the α-amino group
2. Specific for the α-keto acid that accepts the α-amino group.

BIMOLECULAR PING-PONG MECHANISM: The first substrate has to leave before the second one can enter.

1. An amino acid enters, leaves its amino group behind (with the PLP), and leaves as an α-keto acid.
2. A different α-keto acid enters, the reverse reactions occur for it to take the amino group (from the PLP), and it leaves as an amino acid.

IMPORTANT TRANSAMINATION PAIRS

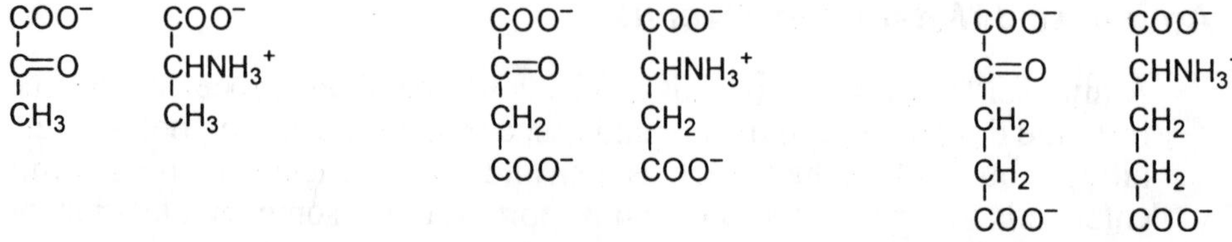

pyruvate/alanine   oxaloacetate/aspartate α-ketoglutarate/glutamate

Note that α-ketoglutarate and oxaloacetate are citric acid cycle intermediates.

DIRECT INCORPORATION OF AMMONIUM IONS INTO ORGANIC MOLECULES

REDUCTIVE AMINATION OF α-KETO ACIDS

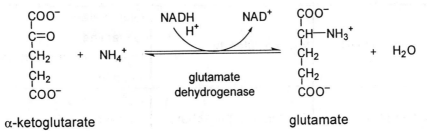

α-ketoglutarate   glutamate

(This reaction typically functions to get rid of NH_4^+, but can also be reversed if excess ammonia is present.)

FORMATION OF THE AMIDES OF ASP AND GLU (TO MAKE ASN AND GLN)

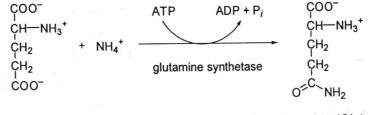

glutamate (Glu)   glutamine (Gln)

The brain uses this reaction to get rid of NH_4^+. Plants couple the reaction above with the following reaction, using two electrons from ferredoxin or NADPH, for a net of 1 Glu per NH_4^+.

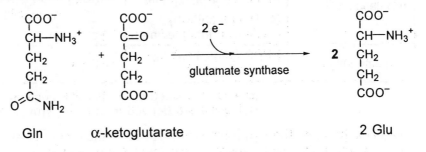

Gln   α-ketoglutarate   2 Glu

Synthesis of the Amino Acids

There's no net synthesis of an amino acid if its α-keto acid precursor isn't independently synthesized by the organism. *de novo* pathways are reaction pathways that synthesize amino acids from metabolic intermediates (not only by transamination).

161

AMINO ACIDS IN THE SAME FAMILY ARE SYNTHESIZED FROM A COMMON PRECURSOR.

Family	Precursor (parent)	Amino Acids
Glutamate family	α-Ketoglutarate	Glutamate
		Glutamine
		Proline (cyclized derivative of Glu)
		Arginine
Serine family	Glycerate-3-phosphate GLUTAMATE	Serine
		Glycine
		Cysteine
Aspartate family	Oxaloacetate (OAA) GLUTAMATE	Aspartate (from OAA + Glu transamination)
		Asparagine (from Asp + Gln transamination)
		Lysine
		Methionine
		Threonine
		Isoleucine*
Pyruvate family	Pyruvate GLUTAMATE	Alanine (from pyruvate + Glu transamination)
		Valine
		Leucine
		Isoleucine*
Aromatic family	Phosphoenolpyruvate Erythrose-4-phosphate (shikimate pathway; chorismate intermediate)	Phenylalanine
		Tyrosine (from chorismate OR from hydroxylation of Phe)
		Tryptophan
Histidine family	PRPP (phosphoribosyl-pyrophosphate)	Histidine

* Isoleucine is listed in two families because the carbons in
isoleucine are derived from both pyruvate and oxaloacetate.

As an exercise, go through figures 14.5–14.12 in the text to be sure you can follow
how a given amino acid is made from each parent molecule. Make your own
diagrams of each family's reactions, and trace specific atoms through the pathways.

BIOSYNTHETIC REACTIONS OF AMINO ACIDS

One-Carbon Metabolism: the transfer of one-carbon groups

ONE-CARBON GROUPS AND THEIR RELATIVE OXIDATION LEVELS:

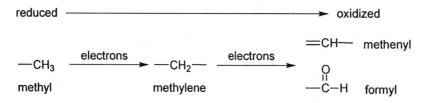

IMPORTANT CARRIERS/DONORS OF ONE-CARBON GROUPS:

1. Tetrahydrofolate (THF) - the biologically active form of folic acid. Once C is attached to THF, the oxidation state can be interconverted (see below). Sources of C groups include Gly, Ser, His, and formate (from Trp).

2. S-Adenosylmethionine (SAM) contains an activated methyl thioether group that transfers its methyl group.

TETRAHYDROFOLATE (THF)

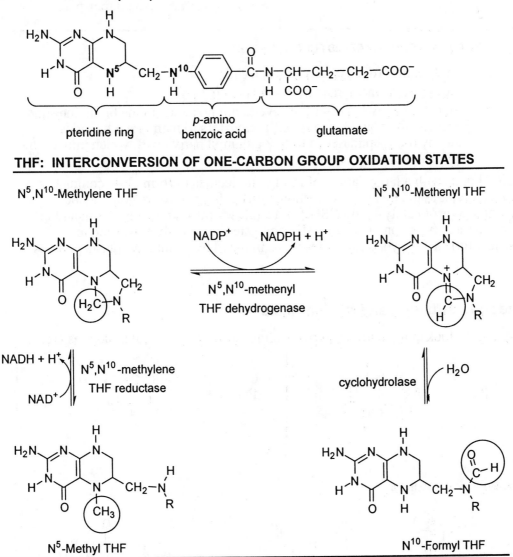

THF: INTERCONVERSION OF ONE-CARBON GROUP OXIDATION STATES

N^5,N^{10}-Methylene THF

N^5,N^{10}-Methenyl THF

NADP⁺ → NADPH + H⁺

N^5,N^{10}-methenyl THF dehydrogenase

NADH + H⁺ / NAD⁺ N^5,N^{10}-methylene THF reductase

cyclohydrolase / H₂O

N^5-Methyl THF

N^{10}-Formyl THF

S-ADENOSYLMETHIONINE (SAM)

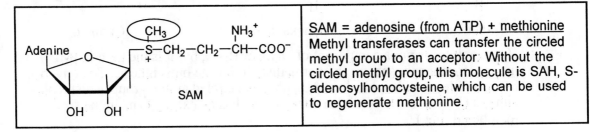

SAM = adenosine (from ATP) + methionine
Methyl transferases can transfer the circled methyl group to an acceptor. Without the circled methyl group, this molecule is SAH, S-adenosylhomocysteine, which can be used to regenerate methionine.

Glutathione: *an important reducing agent formed from Glu, Cys, & Gly*

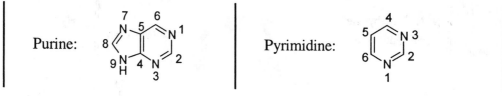

Important functions of GSH (glutathione):

- reduces molecules in various biosyntheses

- protects cells from radiation, O_2 toxicity, environmental toxins

- promotes amino acid transport

GSH TRANSPORT OUT OF CELLS FUNCTIONS TO:

1. transfer sulfur (of cysteine) between cells
2. protect plasma membrane from oxidative damage
3. provide for active transport of several amino acids (in the brain, intestine, pancreas, liver, and kidney): GSH transfers to membrane bound γ-glutamyltranspeptidases to form γ-glutamyl derivatives, which initiates the γ-glutamyl cycle.

GSH bonds with a large variety of foreign molecules to form GSH conjugates, which prepares the toxins for excretion. GSH conjugate formation may be spontaneous or catalyzed by GSH-S-transferases (also known as the ligandins). Before their excretion in urine, GSH conjugates are usually converted to mercapturic acids by a series of reactions initiated by γ-glutamyltranspeptidases.

Nucleotide Structure and Biosynthesis

First, let's look at the purine and pyrimidine bases. The general structures are:

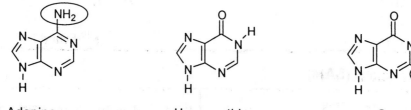

PURINE STRUCTURES

Adenine   Hypoxanthine   Guanine

Hypoxanthine is not a component of DNA or RNA, but it is the base in IMP, inosine-5'-monophosphate, an intermediate in purine biosynthesis. To convert hypoxanthine to either adenine or guanine, the circled amino groups must replace either a C=O or an H. (Note that adenine also has an extra alkene in the six-membered ring.)

PYRIMIDINE STRUCTURES

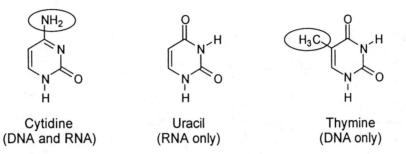

Cytidine	Uracil	Thymine
(DNA and RNA)	(RNA only)	(DNA only)

Uracil monophosphate (UMP) is an intermediate in pyrimidine biosynthesis; molecules containing cytidine and thymine are derived from UMP. Again, the groups that differ from uracil are circled. (Note that cytidine, like adenine, also has an extra double bond in the ring.)

NUCLEOSIDES AND NUCLEOTIDES

Nucleo*S*ide = *S*ugar + Base (purine or pyrimidine)

Nucleo*T*ide = The *T*otal *P*ackage: sugar + base + *P*hosphate, *T*oo.

The carbons in the pentose are numbered beginning with the anomeric carbon, which is consistent with what you learned in Chapter 7. So as not to confuse the pentose carbons with those of the bases, the pentose carbon numbers carry a prime, such as 3' and 5'.

The sugar in nucleotides can either be ribose (for RNA) or deoxyribose (for DNA). The base is connected to the anomeric carbon. The phosphate is connected to the 5' carbon. In deoxyribose, the 2'-OH is replaced by an H.

Since the bases don't have any chiral carbons, the bases can be drawn "facing" the left or right. For example, these two structures of adenine are the same:

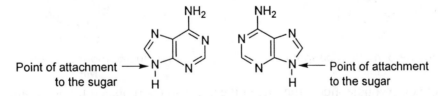

However, once the bases are attached to the pentose, how they are drawn will represent either *syn* or *anti* conformations. Purines occur as both *syn* and *anti*, but pyrimidines are typically *anti* (the base's C=O interferes with the pentose).

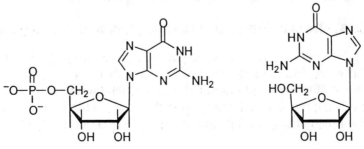

Anti-guanosine-5'-monophosphate   Syn-guanosine-5'-monophosphate

NUCLEOSIDES (RIBOSE + BASE)

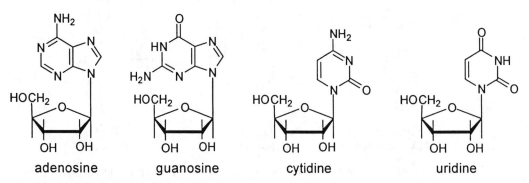

adenosine   guanosine   cytidine   uridine

DEOXYNUCLEOSIDES (DEOXYRIBOSE + BASE)

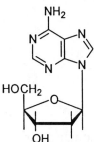
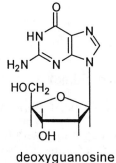
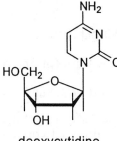
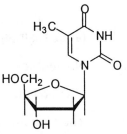

deoxyadenosine   deoxyguanosine   deoxycytidine   deoxythymidine

NUCLEOTIDES (RIBOSE + BASE + PHOSPHATE)

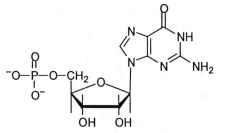

guanosine-5'-monophosphate   cytidine-5'-monophosphate
GMP   CMP

For the deoxynucleotides, the OH at the 2' carbon is replaced by an H, and the abbreviations are designated by a small "d", e.g., dGMP, dCMP.

As an exercise, draw the structures of the nucleotides dAMP, UMP, and dTMP. Check your structures with those in Figure 14.23 (pp. 487-8).

NUCLEOTIDE BIOSYNTHESIS: THE REAL CHALLENGE LIES IN MAKING THE BASE.

Nucleotides can be synthesized in *de novo* pathways (meaning "from scratch") or in "salvage" pathways. Many nucleic acids, such as RNA, are "turned over," meaning they are synthesized and degraded. Rather than having to synthesize the purine nucleotide *de novo*, the cell can recycle the bases. This makes sense when considering how much energy (ATP equivalents) is expended in *de novo* biosynthesis.

166

For *de novo* biosynthesis of...

...purines: the base is built onto ribose-5-phosphate.

...pyrimidines: the base is made first and *then* attached to ribose-5-phosphate.

Where does each atom in the purine or pyrimidine ring come from?

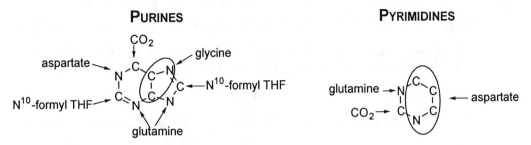

PURINES   **PYRIMIDINES**

STUDY TIP: Write out your own condensed version of Figure 14.24, The Synthesis of Inosine-5'-Monophosphate, and Figure 14.27, Pyrimidine Nucleotide Synthesis. With different colored pens, color the atoms that come from each of the sources shown above. Check your final structure to make sure that it agrees with labels above. This exercise will help immensely to simplify this complex pathway.

PURINES: RINGS ARE BUILT ONTO AN ACTIVATED RIBOSE (PRPP)

PURINE *DE NOVO* PATHWAYS:

1. Ribose-5-phosphate is activated by converting it to PRPP:

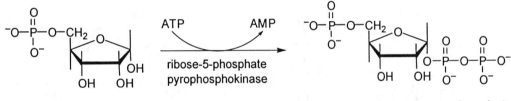

α-D-ribose-5-phosphate
(from the pentose phosphate pathway)

5-phospho-α-D-ribosyl-1-pyrophosphate
(PRPP)

Note that the anomeric carbon of ribose-5-phosphate and PRPP are in the α-configuration. This changes in the next step.

2. PRPP's pyrophosphate group is displaced by the amide nitrogen of glutamine to form 5-phospho-β-D-ribosylamine (and glutamate).

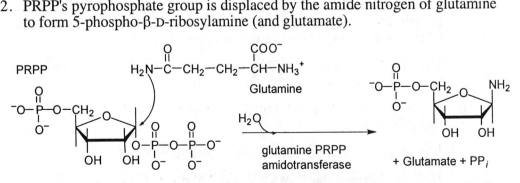

3. Nine further reactions add various groups and close the rings to form IMP. The sources of these groups are listed in the order that they react: Gly, N^{10}-formyl-

THF, Gln, (5-membered-ring closure), CO_2, Asp, (loss of fumarate), N^{10}-formyl-THF, (ring closure to form IMP). Phew!

4. Further reactions change IMP into either AMP or GMP.

PURINE SALVAGE PATHWAYS USE PRPP AND THE FREE BASE → NUCLEOTIDES

Hypoxanthine-guanine phosphoribosyltransferase (HGPRT) catalyzes nucleotide synthesis using PRPP and either hypoxanthine or guanine:

$$\text{Guanine} + \text{PRPP} \rightarrow \text{GMP} + \text{PP}_i$$

$$\text{Adenine} + \text{PRPP} \rightarrow \text{AMP} + \text{PP}_i \quad \text{Adenine phosphoribosyltransferase}$$

PYRIMIDINES: THE BASE IS CREATED FIRST, THEN ATTACHED TO THE SUGAR

1. Carbamoyl phosphate is formed from glutamine and HCO_3^-. This reaction uses one ATP for the phosphate and a second ATP to drive the reaction. It's catalyzed by the cytoplasmic enzyme carbamoyl phosphate synthetase II, a key regulatory enzyme in mammals. It's inhibited by UTP (the product), and stimulated by purine nucleotides (so that the numbers of pyrimidines and purines made are balanced).

2. Carbamoyl phosphate then condenses with aspartate (by aspartate transcarbamoylase).

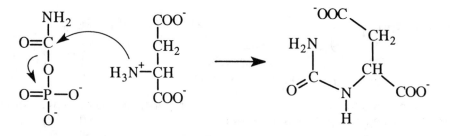

carbamoyl phosphate + aspartate $\rightarrow$ carbamoyl aspartate + P_i

3. The ring is closed by dihydroorotase to form dihydroorotate.

4. Dihydroorotate is then oxidized (hydrogens are removed to form a double bond) by the catalyst dihydroorotate dehydrogenase to form orotate.

5. The ring is attached to PRPP by orotate pyrophosphoribosyl transferase to form orotidine-5'-monophosphate (OMP).

6. A CO_2 is removed to form uridine monophosphate (UMP) (catalyzed by OMP decarboxylase).

UMP serves as a precursor for the other pyrimidine nucleotides.

UMP is phosphorylated twice by ATP to form UTP.

$$\text{UMP} + \text{ATP} \rightarrow \text{UDP} + \text{ADP}$$

$$\text{UDP} + \text{ATP} \rightarrow \text{UTP} + \text{ADP}$$

The amide nitrogen from glutamine replaces the carboxyl group on UTP to form cytidine triphosphate (CTP).

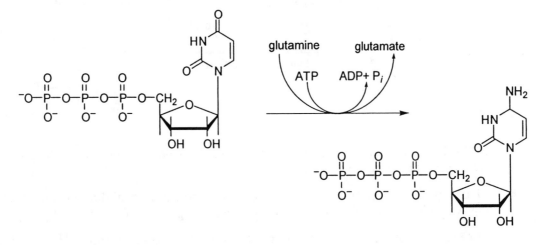

DEOXYRIBONUCLEOTIDES: REDUCTION OF RIBONUCLEOTIDE DIPHOSPHATES

Ribonucleotide reductase replaces the 2'-OH with an H, converting a ribonucleotide *di*phosphate into a deoxyribonucleotide diphosphate.

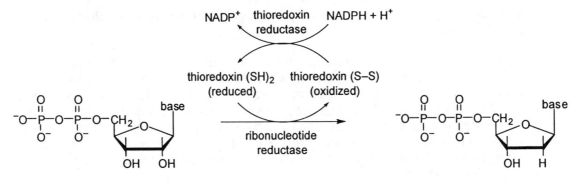

WHAT ABOUT THYMIDINE?

The only nucleotide that's missing is deoxythymidine monophosphate (dTMP).

Note that the only difference between uracil and thymine is a methyl group. The transformation of UMP to dTMP is: UMP → UDP → dUDP → dUMP → dTMP.

Ribonucleotide reductase catalyzes UDP → dUDP. Conversion to dUMP is a simple dephosphorylation. Adding the methyl group to dUMP requires N^5,N^{10}-methylene THF and the enzyme thymidylate synthase to obtain dTMP:

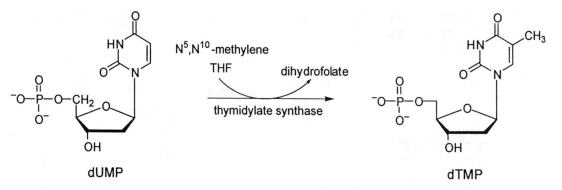

dUMP   dTMP

NADPH donates its electrons (and hydrogens) to the dihydrofolate produced to regenerate THF (tetrahydrofolate).

Since cancer cells grow rapidly, they require large amounts of deoxynucleotides for DNA synthesis. Many anti-tumor or anti-cancer agents block these steps.

For example, methotrexate and aminopterin block the reduction of DHF to THF, and fluorodeoxyuridine monophosphate (FdUMP) inhibits the methylation of dUMP to form dTMP.

Heme: A Complex Fe-Containing Porphyrin Ring Made From Gly & Succinyl-CoA

How are the biosynthesis pathways for heme and chlorophyll similar and how are they different?

The syntheses of heme and chlorophyll are very similar. Although both pathways begin with different starting material, their paths cross when they each form δ–aminolevulinate (ALA). The pathways are identical from ALA to protoporphyrin IX. What distinguishes the remaining pathway is that the protoporphyrin can either be inserted with an Mg^{2+} to form the Mg-protoporphyrin, or it can be inserted with a Fe^{2+} to form protoheme. At this juncture chlorophyll is formed by the addition of a methyl group to the Mg-protoporphyrin and is fully converted by several light-dependent reactions.

CHAPTER 14: ANSWERS TO EVEN-NUMBERED REVIEW QUESTIONS

2.   Transamination reactions represent a method of conserving valuable nitrogen reserves. The reaction is reversible and can be used to convert α-keto acids produced by metabolic reactions to α-amino acids that may be in short supply. Surplus amino acids present in larger amounts than required by current metabolic needs are used as the source of the amino groups.

4.   Nitrogen-fixing organisms solve the problem of oxygen inactivation in several ways. These are: (1) anaerobic organisms live only in anaerobic soil and are not faced with the problem of oxygen inactivation, (2) other organisms physically separate oxygen from the nitrogenase complex. For example many of the cyanobacteria produce specialized nitrogenase-containing cells called heterocysts. The thick cell walls of the heterocysts isolate the enzymes from atmospheric oxygen. In addition, legumes produce an oxygen binding protein called leghemoglobin which traps oxygen before it can interact with the nitrogenase complex.

6.   The Schiff base formed when pyridoxal phosphate reacts with an amino acid loses a proton to form a carbanion. The free electrons of the carbanion are in conjugation with the positively charged nitrogen of the pyridinium ion. As electrons flow to the positively charged nitrogen, the double bond system reorganizes itself to quench the charge on the nitrogen.

8.   Neurotransmitters are either excitatory or inhibitory. Excitatory neurotransmitters (e.g., glutamate and acetylcholine) promote the depolarization of a postsynaptic cell. Inhibitory neurotransmitters (e.g., glycine) inhibit action potentials in postsynaptic cells, i.e., they make the membrane potential more negative.

10.   a.   Alanine belongs to the pyruvate family.
       b.   Phenylalanine belongs to the aromatic family.
       c.   Methionine belongs to the serine family.
       d.   Tryptophan belongs to the aromatic family.
       e.   Histidine belongs to the histidine family.
       f.   Serine belongs to the serine family.

12.   Glutathione is involved in the synthesis of DNA, RNA and the eicosanoids. It is also utilized as a reducing agent which protects cells from radiation and oxygen, and a conjugating agent for environmental toxins. Glutathione is also believed to play a role in amino acid transport.

14.   a. Nucleotide, b. Nucleoside, c. Purine, d. Pyrimidine, and e. Nucleotide (nucleoside triphosphate)

16.   Pyrimidine nucleosides occur predominantly in the *anti* conformation. Steric hindrance between the pentose sugar and the carbonyl oxygen at C-2 of the pyrimidine ring prevents free rotation around the N-glycosidic bond.

18.   Five ATP are required to synthesize a purine by the *de novo* pathway. Only one ATP is required if a purine molecule is recovered by the salvage pathway thus giving a total of four ATP.

20.   Glutamate plays a central role in amino acid metabolism because it and α-ketoglutarate constitute one of the most common α-amino acid/ α-ketoacid pairs used in

transamination reactions. Glutamate also serves as a precursor of several amino acids and as a component of polypeptides. Glutamine serves as the amino group donor in numerous biosynthetic reactions (e.g., purine, pyrimidine and amino sugar synthesis), as a safe storage and transport form of ammonia, and as a component of polypeptides.

22. In ping-pong reactions, the first substrate must leave the active site before the second can enter. In the reaction of alanine with α-ketoglutarate to produce pyruvate and glutamate the following steps take place: (1) the alanine enters the active site and transfers the amino group to pyridoxal phosphate, (2) water enters the reaction site and hydrolyses the Schiff base to produce pyridoxamine phosphate and pyruvate, (3) pyruvate diffuses from the active site, (4) α-ketoglutarate, the second substrate, enters the reaction site and forms a Schiff base with the pyridoxamine phosphate, (5) water hydrolyses the Schiff base to give pyridoxal phosphate and glutamate, and (6) glutamate diffuses out of the active site.

24. The biologically active form of folic acid referred to as tetrahydrofolate or THF is shown below. It is formed by the reduction of folic acid with NADPH, a reaction that is catalyzed by tetrahydrofolate reductase.

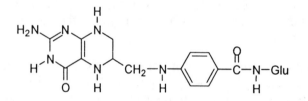

Chapter 14: Answers to Even-Numbered Thought Questions

2. Purine and pyrimidine bases are recycled in salvage pathways. Instead of being degraded to form precursors for catabolic pathways they are, instead, oxidized and excreted as nitrogenous waste products.

4. Glutamate is an excitatory neurotransmitter with stimulating effects on neurons that regulate bodily functions such as blood pressure and body temperature. Individuals who display symptoms after consuming monosodium glutamate apparently possess efficient mechanisms for transporting glutamate across the blood-brain barrier.

6. Radiolabel both the carbon and nitrogen of aspartate and glutamate. If the amino acid is used in the ring assembly, then both carbon and nitrogen should bear the label. If nitrogen exchange takes place, only the carbon atom will be labeled. In addition, isolating each intermediate allows the origin of each atom to be traced.

8. Arginine is normally synthesized by the urea cycle. In small children, the urea cycle is not fully functional. Consequently, arginine must be obtained from external sources.

10. Refer to Figure 14.27, p. 494: Pyrimidine Nucleotide Synthesis.

---

zein - a protein found in corn, also, a great (and official) Scrabble word!

CHAPTER FIFTEEN

Nitrogen Metabolism II: Degradation

HOW DO VARIOUS ANIMAL SPECIES DISPOSE OF NITROGENOUS WASTE?

The nitrogen in amino acids is removed by deamination reactions (transamination and oxidative deamination) and converted to ammonia, which must be detoxified and/or excreted as fast as it's made. The form in which nitrogen is excreted depends on the availability of water.

AMMONOTELIC: many aquatic animals that can excrete ammonia, which dissolves in the surrounding water and is quickly diluted.

UREOTELIC: terrestrial animals (such as mammals) that convert ammonia into urea because they must conserve body water. Urea can be excreted without a large loss of water.

URICOTELIC: animals such as birds, certain reptiles, and insects, that convert ammonia to uric acid because they have more stringent water conservation requirements. In many animals, uric acid is also the nitrogenous waste product of purine nucleotide catabolism.

AMINO ACID CATABOLISM CAN BE DIVIDED INTO THREE STAGES:

1. deamination (removal of the α-amino group and transport to the liver)

2. urea synthesis (to excrete nitrogen; occurs only in the liver)

3. conversion of the carbon skeleton to one of seven metabolic intermediates

Deamination: Transamination and Oxidative Deamination

The first step in amino acid catabolism is almost always to remove the α-amino group by transamination, in which the α-amino group is transferred to α-ketoglutarate to form glutamate (Glu).

TRANSAMINATION:

173

Excess NH₄⁺ Is Carried to the Liver by Glutamine or Alanine: [1]

Muscle (Alanine Cycle):

("α-kg" = α-ketoglutarate)

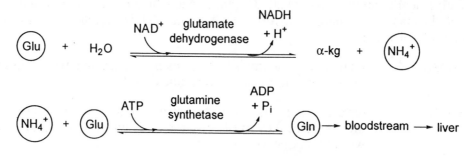

Most Other Tissues: Oxidative Deamination,[2] Then Glutamine Synthetase

In most tissues (except the liver), two reactions (and two separate glutamates) are needed to transport NH_4^+ through the blood in the form of glutamine (Gln).

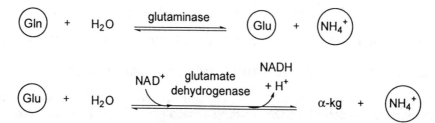

Liver: Frees NH₄⁺ From Gln (From Other Tissues) and Ala (From Muscle)

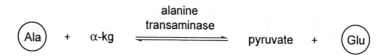

Gln loses its NH_4^+ to regenerate Glu (in a reaction catalyzed by glutaminase). Glu is then oxidatively deaminated to release NH_4^+.

Other NH₄⁺ -Generating Reactions

1. Amino acid oxidases (in the liver and kidney)
2. Serine and threonine dehydratases (because serine and threonine can't be transaminated): Ser → pyruvate; Thr → α-ketobutyrate
3. Bacterial urease

---

[1] Why doesn't glutamate carry nitrogen to the liver? One possible reason is that glutamate is also a neurotransmitter in the brain. If glutamate levels in the blood are elevated, this could cause the brain to "short circuit." In fact, this is an explanation for the headaches that some people experience as a result of consuming MSG (monosodium glutamate), an seasoning ingredient sometimes found in Asian cuisine.

[2] We saw the reverse of this reaction, reductive amination, in Chapter 14.

THE UREA CYCLE OCCURS IN LIVER CELLS

THE OVERALL REACTION OF UREA SYNTHESIS:

$$CO_2 + NH_4^+ + Asp + 3\ ATP + 2\ H_2O \rightarrow$$

$$urea + fumarate + 2\ ADP + 2\ P_i + AMP + PP_i + 5\ H^+$$

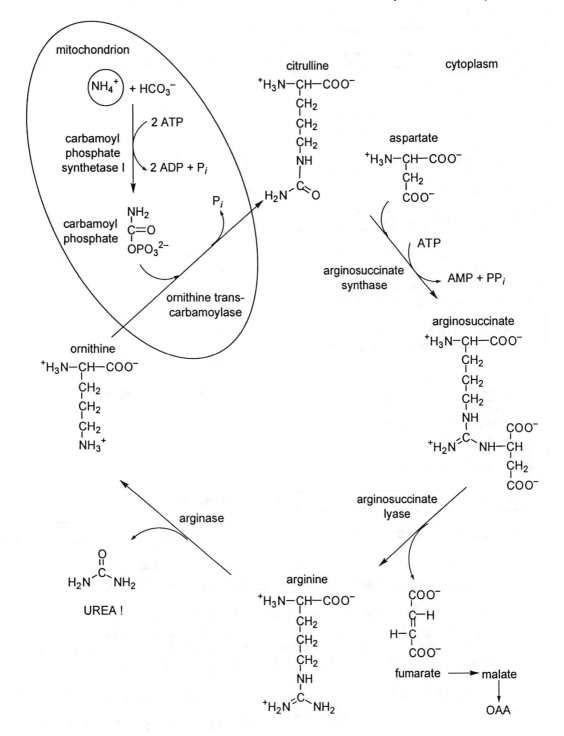

THE UREA CYCLE: REACTION NOTES

1. Carbamoyl phosphate synthetase I	$NH_4^+ + HCO_3^- + 2\ ATP \rightarrow$
Mitochondrial matrix Similar to a reaction that we saw in pyrimidine biosynthesis Uses 2 ATP: one to activate HCO_3^-, one to phosphorylate carbamate	$\qquad$ carbamoyl phosphate + 2 ADP + P_i + 3H$^+$
2. Ornithine transcarbamoylase	carbamoyl phosphate + ornithine $\rightarrow$ citrulline + P_i
Mitochondrial matrix Carbamoyl phosphate has a high phosphate group transfer potential. Losing its P_i drives this reaction forward. Ornithine is an amino acid; in fact, it looks just like lysine but with one less -CH$_2$-group.	citrulline $\rightarrow$ transport to cytoplasm
3. Arginosuccinate synthase	citrulline + Asp $\rightarrow$ arginosuccinate
Cytoplasm Cleavage of PP$_i$ (by pyrophosphatase) pulls this reversible reaction forward. The α-amino of Asp adds the second N of urea Uses 1 ATP, but since AMP, not ADP, is formed, it's equivalent to using two ATP.	
4. Arginosuccinate lyase	arginosuccinate $\rightarrow$ Arg + fumarate
Cytoplasm Asp's N is left behind when fumarate is cleaved from argininosuccinate to form Arg. Fumarate's carbons came from Asp.	
5. Arginase	Arg + H$_2$O $\rightarrow$ ornithine + urea
Cytoplasm Hydrolyzes Arg to urea + ornithine	ornithine $\rightarrow$ transport back to mitochondrion urea $\rightarrow$ transport via bloodstream to kidneys $\rightarrow$ excretion

KREBS BICYCLE: THE UREA CYCLE – CITRIC ACID CYCLE CONNECTION

What happens to the fumarate? Where can those four carbons go? Fumarate supplied by the urea cycle can enter the mitochondrion and enter the citric acid cycle: fumarate $\rightarrow$ malate $\rightarrow$ oxaloacetate (OAA). From OAA, three paths are possible, depending (of course) on the metabolic needs of the cell.

- Transamination $\quad$ OAA + α-amino acid $\rightarrow$ α-keto acid + Asp $\rightarrow$ urea cycle

- Gluconeogenesis: OAA $\rightarrow$ PEP $\rightarrow$etc. $\rightarrow$ glucose-6-phosphate $\rightarrow$ glucose

- Citric Acid Cycle: OAA + acetyl-CoA $\rightarrow$ citrate $\rightarrow$ etc.$\rightarrow$ energy

Control of the Urea Cycle

Short term control: substrate concentrations. So, it makes sense that urea synthesis is stimulated when the diet is high in protein or during starvation (when muscle protein is degraded for energy). At these times, there would be more NH_4^+ that needs to be excreted.

Carbamoyl phosphate synthetase I: allosterically activated by N-acetylglutamate, an indicator of Glu concentration (Glu + acetyl-CoA → N-acetylglutamate)

Catabolism of Amino Acid Carbon Skeletons to Metabolic Intermediates

The carbon skeletons are eventually degraded to one of seven intermediates. The degradation of the amino acid carbon skeletons are classified in terms of their converted end products. Acetyl-CoA and acetoacetyl-CoA are grouped together.

KETOGENIC AMINO ACIDS are degraded to form acetyl-CoA or acetoacetyl-CoA, which can be converted to either fatty acids or ketone bodies. Remember that no net synthesis of glucose can occur with acetyl-CoA or acetoacetyl-CoA.

GLUCOGENIC AMINO ACIDS are degraded to form intermediates that can be used in gluconeogenesis. The intermediates are pyruvate, α-ketoglutarate, succinyl CoA, fumarate, and oxaloacetate. These intermediates can then be catabolized by the citric acid cycle or converted to fatty acids, ketone bodies, or glucose.

Keep in mind the following reactions that can result in some overlap or confusion between groups:

pyruvate → acetyl-CoA
acetoacetate → acetoacetyl-CoA → acetyl-CoA
citric acid cycle reactions

Acetyl-CoA	Lys, Trp, Tyr, Phe, Leu	Acetoacetyl-CoA is an intermediate common to all five amino acids. acetoacetyl-CoA → acetyl-CoA Phe→Tyr→acetoacetate→acetoacetyl-CoA
Pyruvate	Ala, Cys, Thr, Gly, Ser	Thr→Gly→Ser→pyruvate (Acetyl-CoA is not an intermediate between these five amino acids and pyruvate.)
α-Ketoglutarate	Gln, Arg, Pro, His, Glu	Glu is an intermediate common to the other four amino acids. Glu → α-ketoglutarate
Succinyl CoA	Met, Ile, Val, Thr	Propionyl-CoA is an intermediate common to all four amino acids.
Oxaloacetate	Asp, Asn	
Fumarate	Tyr	Catabolism of Tyr forms both acetoacetate and fumarate.

WHAT ROLE DOES PROTEIN TURNOVER PLAY IN CELLULAR METABOLISM?

Protein turnover is the continuous degradation and re-synthesis of proteins. Metabolic flexibility is afforded by relatively quick changes in the concentrations of key regulatory enzymes, peptide hormones, and receptor molecules. Protein turnover also protects cells from accumulation of abnormal or damaged proteins. Finally, the process of organismal growth and development are as dependent on timely degradative reactions as they are on synthetic ones.

HOW ARE PROTEINS TARGETED FOR DEGRADATION?

The mechanisms by which proteins are targeted for destruction by ubiquitination or other degradative processes are not fully understood. However, the following features of proteins appear to signal their destruction:

a. <u>N-terminal residues</u>: The N-terminal residue of a protein is partially responsible for its susceptibility to degradation. For example, proteins with methionine or alanine N-terminal residues have substantially longer half-lives than do those with leucine or lysine.

b. <u>PEST sequences</u>. Proteins which have extended sequences containing proline, glutamate, serine, and threonine have been observed to possess half-lives of less than two hours.

c. <u>Oxidized residues</u>. Oxidized amino acid residues (i.e., residues which are altered by oxidases or attacked by ROS) promote protein degradation.

DEGRADATION OF SELECTED NEUROTRANSMITTERS

WHAT ROLE DOES THE DESTRUCTION OF NEUROTRANSMITTERS PLAY IN THE FUNCTIONING OF NEURONS AND MUSCLE CELLS?

Neurotransmitter degradation in a timely and efficient fashion is required for maintaining precise information transfer in the nervous system. Degradation of neurotransmitters essentially results in the inactivation of the signal.

Acetylcholine: inactivated by acetylcholinesterase which hydrolyzes acetylcholine to acetate and choline

Catecholamines (epinephrine, norepinephrine, dopamine): inactivated by transport out of the synaptic cleft followed by monoamine oxidase (MAO) oxidation; also inactivated by methylations by catechol-O-methyltransferase (COMT); epiniphrine is catabolized in nonneural tissue

Serotonin: after reuptake into nerve cells, MAO oxidation followed by aldehyde dehydrogenase oxidation

NUCLEOTIDE CATABOLISM

Digestion:

nucleic acids + H_2O → oligonucleotides   nucleases (DNases, RNases)

oligonucleotides + H_2O → nucleotides   phosphodiesterases

nucleotide + H_2O → nucleoside + P_i nucleotidases

nucleoside + P_i → base + ribose-1-phosphate   nucleosidases

Purines are degraded to uric acid.

Pyrimidines are degraded to NH_4^+, CO_2, and either β-alanine or β-aminoisobutyrate.

Purine Catabolism

For AMP, GMP, IMP, and XMP, the first step is the same: dephosphorylation by 5'-nucleotidase to form the nucleosides adenosine, guanosine, inosine, and xanthosine, respectively.

Adenosine is deaminated to form inosine, then the ribose is removed. (Note that adenine is not an intermediate.) AMP may also be deaminated first (to form IMP) then dephosphorylated (to form inosine). Inosine's ribose is removed to form the free base hypoxanthine, which is then oxidized to xanthine.

Guanosine loses its ribose to form the free base guanine, which is deaminated to form xanthine.

Finally, xanthine is oxidized to uric acid by xanthine oxidase.

Animals other than primates can degrade uric acid into other nitrogen products such as urea and NH_4^+ .

The structures are outlined on the following page. Look at the structure of uric acid. Note that the purine rings remain intact.

Pyrimidine Catabolism

As an exercise, compare the outline of pyrimidine catabolism (following purine catabolism) with Figure 15.14 in your text (p. 524). Write in the groups that are added and removed (NH_4^+, P_i) and the names of the enzymes (or, recopy this figure onto a separate sheet of paper). Drawing careful comparisons between the molecular structures will help you to understand and learn this pathway. Note that the pyrimidine ring can be degraded, whereas the purine ring could not.

Purine Catabolism

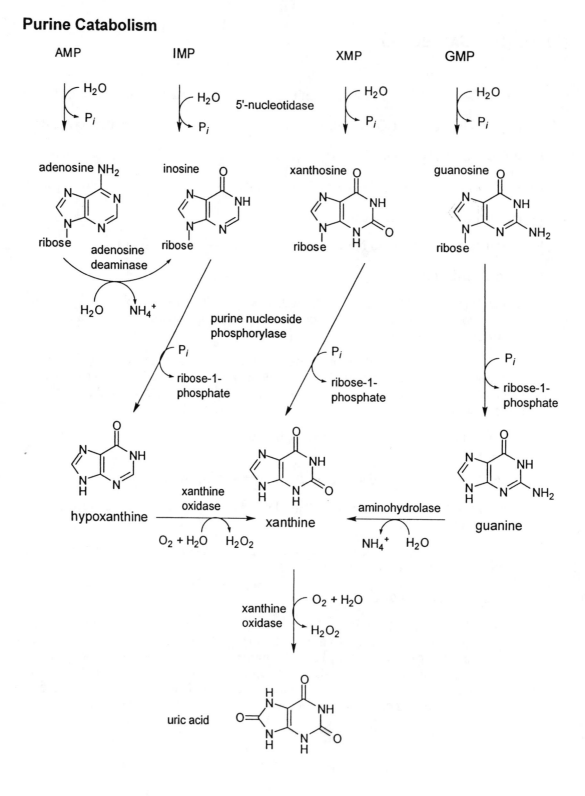

Pyrimidine Catabolism

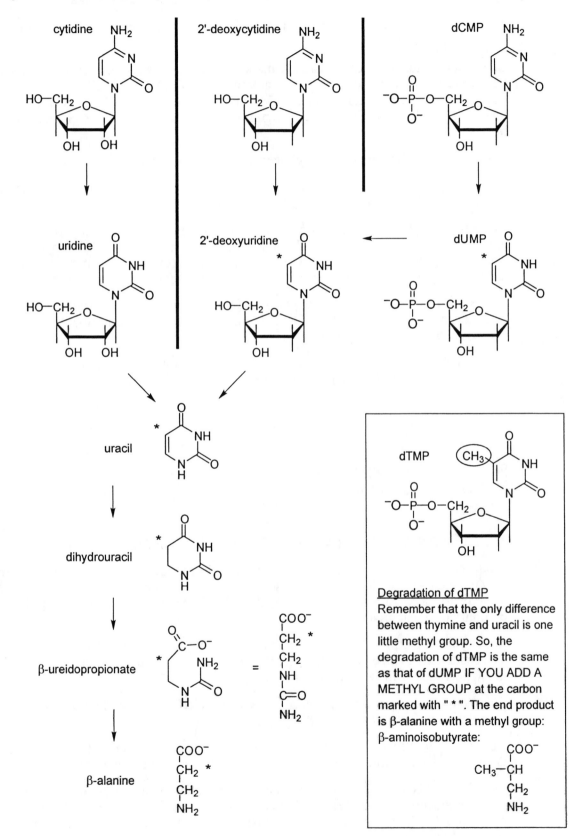

cytidine

2'-deoxycytidine

dCMP

uridine

2'-deoxyuridine

dUMP

uracil

dihydrouracil

β-ureidopropionate

β-alanine

dTMP

<u>Degradation of dTMP</u>
Remember that the only difference between thymine and uracil is one little methyl group. So, the degradation of dTMP is the same as that of dUMP IF YOU ADD A METHYL GROUP at the carbon marked with " * ". The end product is β-alanine with a methyl group: β-aminoisobutyrate:

CHAPTER 15: ANSWERS TO EVEN-NUMBERED REVIEW QUESTIONS

2.   The major nitrogen-containing excretory molecules are ammonia, urea, uric acid, allantoin, and allantoate.

4.   The structural features that apparently mark proteins for destruction are: (1) certain N-terminal amino acid residues (e.g., methionine or alanine), (2) Peptide motif sequences (e.g., amino acid sequences with proline, glutamic acid, serine and threonine), and (3) oxidized residues (amino acid residues whose side chains have been oxidized by oxidases or ROS).

6.   a. Ketogenic, b. Ketogenic, c. Glycogenic, d. Glycogenic, e. Glycogenic, f. Both.

8.   Purine rings are degraded to uric acid. A significant percentage of uric acid is excreted in the urine. (Refer to Figure 15.12.)

10.   In the muscle, pyruvate undergoes a transamination reaction and is converted to alanine. Alanine is then transferred to the liver, where it is reconverted to pyruvate. The amino group is transferred to α-ketoglutarate thus forming glutamate, which is subsequently oxidatively deaminated to form α-ketoglutarate and ammonia.

12.   The NADH formed during the conversion of fumarate to aspartate results in the synthesis of approximately 2.5 ATP. Therefore, the net ATP requirement for urea synthesis is approximately 1.5 moles of ATP per mole of urea.

14.   The branched-chain amino acids (leucine, isoleucine, and valine) are metabolized in tissues, such as muscle, where they are principally used to synthesize nonessential amino acids.

16.   a. uric acid - birds, reptiles and insects

      b. urea – mammals

      c. allantoate - bony fish

      d. NH_4^+ - aquatic animals

      e. allantoin – some mammals

CHAPTER 15: ANSWERS TO EVEN-NUMBERED THOUGHT QUESTIONS

2.   As the concentration of glutamate (as well as its deamination product ammonia) rises, the enzyme N-acetylglutamate synthase catalyzes the synthesis of N-acetylglutamate, an activator of carbamoyl phosphate synthetase I. The latter enzyme catalyzes the first committed step in urea synthesis.

4.   Tetrahydrobiopterin is a cofactor in the oxidation of phenylalanine to form tyrosine. The sustained absence of this cofactor would result in a buildup of phenylalanine and the appearance of the symptoms of PKU.

6.   These amino acids are intermediates in the urea cycle. Therefore, their addition stimulates the formation of urea.

8.   Because of the structural similarities to purine, caffeine is converted to a variety of derivatives by xanthine oxidase (e.g., 1-methyluric acid and 7-methylsxanthine).

10.   In addition to being less toxic than ammonia, urea and uric acid (the nitrogenous waste products of terrestrial animals) require significantly less water for their excretion. Recall that ammonotelic organisms excrete ammonia directly into the surrounding water.

Integration of Metabolism

OVERVIEW OF METABOLISM: CONTROLLING THE BALANCE OF ANABOLIC AND CATABOLIC PROCESSES

Coordination of Metabolism by the Nervous and Endocrine Systems

NERVOUS SYSTEM: Axons release neurotransmitters into synapses (intercellular spaces). Neurotransmitters bind to nearby target cells and evoke a specific response from those cells.

ENDOCRINE SYSTEM:

HORMONES (chemical signals) are secreted into the bloodstream, which transports them to target cells. Hormones bind to specific receptor molecules (often on the target cell's surface) and triggers a response inside the cell. Most hormone-induced changes alter the activity or concentration of enzymes.

SECOND MESSENGERS may be released inside the cell as a result of hormonal binding to a receptor on the surface. (Second messengers are described later.) Second messengers that bind to an enzyme may cause an amplification of the signal called an enzyme cascade.

STEROID HORMONES (lipid-soluble) diffuse into a target cell and binds to a receptor protein in the cytoplasm. The hormone-receptor complex moves to the nucleus and binds to specific sites on DNA, altering a cell's pattern and rate of gene transcription (and protein synthesis).

THE DIVISION OF LABOR

Small Intestine: aids in the digestion and absorption of nutrients such as carbohydrates, lipids, and proteins

Enterocytes of the small intestine absorb nutrients, then transport them into the blood and lymph. Glutamine supplies most of the enterocyte's energy needed for active transport and lipoprotein synthesis.

Liver: plays a key role in carbohydrate, lipid, and amino acid metabolism:

- monitors and regulates the chemical composition of blood and synthesizes several plasma proteins.
- distributes several types of nutrients to other parts of the body
- responsible for reducing fluctuations in nutrient availability caused by drastic dietary changes (such as intermittent feeding and fasting, or high-protein vs. high-carbohydrate diets)
- processes foreign molecules (a critically important protective role)

Muscle

The energy sources which are used to provide ATP for muscle contraction depend on the degree of muscular activity and the physical status of the individual concerned. During fasting and prolonged starvation, some skeletal muscle protein is degraded to provide amino acids to the liver for gluconeogenesis. The cardiac muscle in the heart must continuously contract to sustain blood flow throughout the body. To maintain its continuous operation, cardiac muscle relies mainly on fatty acids.

Adipose Tissue stores energy in the form of triacylglycerols...

...or degrades fat stores to generate energy-rich fatty acids and glycerol, depending on whether nutrients are in excess or whether ATP synthesis is needed.

Brain ultimately directs most metabolic processes in the body

Much of the body's hormonal activity is controlled either directly or indirectly by the hypothalamus and the pituitary gland. Under normal conditions the brain uses glucose as its sole fuel. Under conditions of prolonged starvation, the brain can adapt by using ketone bodies as an energy source.

Kidney has several important functions:

- filtration of blood plasma to excrete water-soluble waste products,
- reabsorption of electrolytes, sugars, and amino acids from the filtrate,
- regulation of blood pH, and
- regulation of the body's water content.

Energy needed for transport processes is provided largely by fatty acids and glucose. Under normal conditions the small amounts of glucose which are formed by gluconeogenesis are used only within certain kidney cells. The rate of gluconeogenesis increases during starvation and acidosis.

THE FEEDING-FASTING CYCLE

Consuming food intermittently is possible because of elaborate mechanisms for storing and mobilizing energy-rich molecules derived from food. Regulation of opposing pathways ensure that they don't occur simultaneously.

FEEDING-FASTING SUMMARY: EFFECTS OF INSULIN AND GLUCAGON

	Postprandial state after a meal elevated blood nutrient levels	Postabsorptive state after a fast low blood nutrient levels
	Insulin	**Glucagon**
Blood	Lowers blood glucose	Raises blood glucose
Liver	glucose → glycogen	glycogen → glucose
Muscle	amino acids → protein	protein → amino acids
Adipose tissue	fatty acids, glycerol → fats	fats → fatty acids, glycerol

The Feeding Phase: food is consumed, digested, and absorbed...

Absorbed nutrients are then transported to various organs where they are either used or stored. During the feeding phase, hormones such as gastrin, secretin, and cholecystokinin stimulate the secretion of various digestive enzymes or aids such as bicarbonate and bile.

LIPIDS: Chylomicrons transport lipid molecules from the small intestine, through the lymph and the bloodstream to target tissues. Chylomicron remnants (chylomicrons after most triacylglycerols have been removed) are taken up by the liver where they're reused or degraded. Also, elevated fatty acids in the blood promotes lipogenesis in adipose tissue.

GLUCOSE: As glucose moves from the small intestine to the liver via the blood, β-cells within the pancreas are stimulated to release insulin. The release of insulin triggers several processes which ensure the storage of nutrients. In addition, insulin also influences amino acid metabolism. For example, insulin promotes the transport of amino acids into cells and stimulates protein synthesis (in most tissues).

The Fasting Phase

EARLY POSTABORPTIVE STATE: glucagon is released as blood glucose and insulin levels return to normal. Glucagon prevents hypoglycemia by promoting glycogenolysis and gluconeogenesis in the liver.

PROLONGED FAST: Several strategies used to maintain blood glucose levels include:
- norepinephrine stimulates lipolysis in adipose tissue (releases fatty acids that provide an alternative energy source)
- glucagon increases gluconeogenesis, using amino acids from muscle

STARVATION: Metabolic changes; fatty acids from adipose, ketone bodies from liver

EXERCISE AND NUTRIENT METABOLISM (SPECIAL INTEREST BOX 16.1)

BASAL METABOLIC RATE (BMR)

AEROBIC EXERCISE: The citric acid cycle is zooming! Muscle uses both glucose and fatty acids as fuel. Epinephrine and norepinephrine are released, causing more fatty acids to be released from adipose tissue. After muscle and liver glycogen stores are depleted, the muscle's capacity to generate energy drops to 60% (remember that the citric acid cycle needs some glucose to supply the oxaloacetate).

ANAEROBIC EXERCISE: Glycolysis supplies ATP until lactate levels are too high.

ENDURANCE TRAINING is aerobic exercise that uses large amounts of muscle mass for a minimum amount of time, nonstop, at the training heart rate [= 65% of the maximum heart rate, which is 220 − (person's age)], and is performed regularly.

EFFECTS OF ENDURANCE TRAINING include: increase number of mitochondria per muscle fiber; increased efficiency of fatty acid degradation (resulting from increased synthesis of enzymes, transport proteins, and other molecules); and increased number of capillaries in muscle tissue, resulting in a more efficient exchange of nutrients and waste products.

INTERCELLULAR COMMUNICATION: HORMONES, GROWTH FACTORS, AND SECOND MESSENGERS

In mammals, most metabolic activities are controlled to a certain extent by hormones. Hormones can have specific effects on one type of target cell or on a variety of target cells. Some hormones have different effects on different target cells.

Endocrine hormones vs. paracrine hormones vs. steroid hormones

The Hormone Cascade System

To ensure proper control of metabolism, the synthesis and secretion of many hormones are regulated by a complex cascade mechanism that is ultimately controlled by the central nervous system. Various types of sensory signals are received by the hypothalamus, an area in the brain that links the nervous and endocrine systems. Once it is appropriately stimulated, the hypothalamus induces the secretion of several hormones produced by the anterior lobe of the pituitary gland.

HOW ARE THE STRUCTURES OF THE HYPOTHALAMUS AND PITUITARY GLAND RELATED TO THE FUNCTIONING OF THE PROCESSES THEY REGULATE?

The pituitary gland, which is attached to the hypothalamus by the pituitary stalk, consists of two distinct parts: the anterior lobe, or adenohypophysis, and a posterior lobe, or neurohypophysis. The hypothalamus synthesizes a series of specific peptide-releasing hormones. Hormones then pass into a capillary bed referred to as the hypothalamohypophyseal portal system, which transports them directly to the adenohypophysis. These peptides stimulate specific cells to synthesize and secrete one or more types of hormone. The hormones of the anterior pituitary are sometimes referred to as tropic, since they both stimulate the synthesis and release of hormones from other endocrine glands.

The anatomy and function of the posterior pituitary differ from those of the anterior lobe. The hormones secreted by the neurohypophysis are actually synthesized in separate types of neurons which originate in the hypothalamus.

MECHANISMS THAT PROTECT AGAINST OVERSTIMULATION BY HORMONES

Desensitization: target cells decrease the number of cell surface receptors (or inactivate receptors) in response to changes in stimulation

Down-regulation - the reduction in cell surface receptors in response to stimulation by specific hormone molecules. In down-regulation, receptors are internalized by endocytosis. Depending on the cell type and several other factors, the receptors may eventually be recycled back to the cell surface or degraded.

Insulin-resistance - diabetes caused by a decrease in functional insulin receptors

Growth Factors (cytokines): hormone-like polypeptides and proteins that control cell growth, division, and differentiation:

EPIDERMAL GROWTH FACTOR (EGF) - a mitogen (stimulates cell division) for epidermal and gastrointestinal lining cells; triggers cell division when it binds to plasma membrane EGF receptors (trans-membrane tyrosine kinases)

PLATELET-DERIVED GROWTH FACTOR (PDGF) - secreted by blood platelets during clotting; stimulates mitosis in fibroblasts and other cells during wound healing

SOMATOMEDINS - secreted by the liver into the bloodstream; mediate the growth-promoting actions of GH; include insulinlike growth factors I and II (IGF-I, IGF-II); bind to cell surface receptors that are also tyrosine kinases

INTERLEUKIN-2 (IL-2) - secreted by T cells after they've been activated by binding to a specific antigen-presenting cell; stimulates cell division so that numerous identical T cells are produced; regulates the immune system, promotes cell growth and differentiation

INTERFERONS - growth inhibitors- Type I: protects cells from viral infection; Type II: inhibits the growth of cancer cells, also has several immunoregulatory effects

TUMOR NECROSIS FACTORS (TNF): growth inhibitors that are toxic to tumor cells; suppress cell division

MECHANISMS OF HORMONE ACTION

WATER-SOLUBLE HORMONES are secreted into the bloodstream, transported to target cells, and bind to specific receptor molecules (often on the target cell's surface). Binding triggers a mechanism that initiates a phosphorylation cascade either directly or via a second messenger.

Second Messengers: cAMP, cGMP, DAG, IP$_3$, Ca^{2+}, and the inositol phospholipid system

Signal transduction - the amplification of a hormonal message by a second messenger. Once generated, the signal must be terminated rapidly.

cAMP - GENERATED FROM ATP BY ADENYLATE CYCLASE

G-proteins mediate the interaction between the cAMP receptor and adenylate cyclase.

cAMP molecules diffuse into cytoplasm, then bind to and activate cAMP-dependent protein kinase that phosphorylates and alters the activity of key regulatory enzymes.

cGMP - SYNTHESIZED FROM GTP BY GUANYLATE CYCLASE (MEMBRANE-BOUND OR CYTOPLASMIC)

ANF (atrial natriuretic factor) activates membrane-bound guanylate cyclase, producing cGMP, which activates the phosphorylating enzyme protein kinase G; lowers blood pressure

enterotoxin binds to and activates another type of membrane-bound guanylate cyclase; causes diarrhea

THE PHOSPHATIDYLINOSITOL CYCLE AND CALCIUM MEDIATES THE ACTION OF HORMONES AND GROWTH FACTORS

(See Figure 16.12, page 554 in your text.)

DAG (DIACYLGLYCEROL) ACTIVATES PROTEIN KINASE C

Protein kinase C phosphorylates specific regulatory enzymes

IP$_3$ (INOSITOL-1,4,5-TRIPHOSPHATE): ITS RECEPTOR IS A CALCIUM CHANNEL IN THE SER.

When activated, the channel opens and calcium ions flow through, and the action of Ca^{2+}-regulated proteins are affected.

Steroid and Thyroid Hormones Switch Genes On or Off, changing the pattern of proteins that an affected cell makes

LIPID-SOLUBLE STEROID HORMONES require transport proteins to travel through the bloodstream.

At the target cell, a steroid hormone:

- dissociates from its transport protein
- diffuses through the plasma membrane
- binds to a receptor protein in the cytoplasm or in the nucleus (If in the cytoplasm, the hormone-receptor complex moves to the nucleus.)
- in the nucleus: the hormone-receptor complex binds to the base sequence of and HRE (hormone response elements = specific sites on DNA) via zinc finger domains
- transcription of a specific gene is altered; this affects protein synthesis

Since several HREs can bind to the same hormone-receptor complex, the expression of numerous genes can be altered simultaneously.

The Insulin Receptor: a transmembrane glycoprotein with 4 subunits

Two α-subunits extend out of the cell and bind insulin

Two β-subunits extend through the membrane and contain a tyrosine kinase domain

When insulin binds to the receptor:

- activates tyrosine kinase activity, causing a phosphorylation cascade
- inhibits hormone-sensitive lipase in adipocytes
- activates a phosphatase that removes a phosphate from the lipase
- induces transfer to the cell's surface of several types of protein that affect the uptake of nutrients (*Examples* of proteins that migrate to the surface: glucose transporter, receptors for LDL and IGF-II)

SPECIAL INTEREST BOX 16.2: HORMONE-RELATED DISEASES

WHICH DISEASES RESULT FROM THE OVERPRODUCTION OR UNDERPRODUCTION OF HORMONES?

The oversecretion of hormone molecules is most often caused by a tumor. Several types of pituitary tumors cause endocrine disease.

- Cushing's disease is a condition characterized by obesity, hypertension and elevated blood glucose levels.

- Gigantism, in which there is a pronounced growth of long bones, is caused by excessive secretion of growth hormone during childhood.

- In adulthood, excessive growth hormone production causes acromegaly, a condition in which connective tissue proliferation and bone thickening result in coarsened and exaggerated facial features, as well as enlarged hands and feet.

However, not all hypersecretion diseases are caused by tumors. For example, Graves disease, the most common type of hyperthyroidism, is an autoimmune disease.

Inadequate hormone production has a variety of causes. The most common of these are the autoimmune destruction of hormone-producing cells, genetic defects, or an inadequate supply of precursor molecules.

- Addison's disease is a disorder in which there is inadequate adrenal cortex function. The most common cause of Addison's disease is the autoimmune destruction of the adrenal gland.

- Hypothyroidism may be the result of autoimmune disease (Hashimoto's disease) or the deficient synthesis of thyroid hormones such as TSH and TRH. Because adequate ingestion of iodine is a prerequisite for thyroid hormone synthesis, iodine deficiency can cause hypothyroidism. In children, a thyroid hormone deficiency called cretinism causes depressed growth and mental retardation. In adults, myxedema, also a thyroid hormone deficiency, results in symptoms such as edema and goiter.

- Growth hormone deficiency may be hereditary or a consequence of a pituitary tumor or head trauma. Congenital growth hormone deficiency results in shortened stature known as dwarfism.

- Diabetes insipidus is a malady characterized by copious amounts of very diluted urine. It is caused by either an inadequate synthesis of vasopressin or the failure of the kidney to respond to vasopressin.

SPECIAL INTEREST BOX 16.3: DIABETES MELLITUS

In insulin-dependent diabetes mellitus, also called type I diabetes, inadequate amounts of insulin are secreted because of the destruction of the β-cells in the pancreas. Type I diabetes can also occur because of a mutated or truncated form of insulin. Type I diabetic individuals do not have any functional insulin and are usually thin, since glucagon doesn't have a competitor, and all food ingested is immediately converted to energy and used (instead of being stored).

Noninsulin-dependent diabetes mellitus, also called type II or adult onset diabetes, is caused by a loss of responsiveness of the target tissue to insulin. Type II onset diabetes is caused because of the continuous stimulation of the receptor (in obese people); as a conclusion the individuals need more and more insulin to satisfy the needs of the receptor. In other words, the individuals become resistant to insulin. However, the continuous production of insulin (because of a continuous stimulation) has a bad effect on other tissues, like the adipose tissue (where fat storage increases) and on the organism as a whole (individuals are continuously hungry!!!).

Type II diabetes individuals have too much insulin. While this large amount of insulin has little or no effect on the storage of glucose in the form of glycogen, it does have an effect on other tissues, resulting in individuals who tend to be overweight because of problems with energy storage.

NUTRITION AND CATABOLISM

The energy content of food is measured as calories. Calories are determine by the amount of heat given off when a food substance is completely oxidized. The caloric value of food is slightly reduced in our bodies due to incomplete digestion and metabolism. The general caloric content of various foods is:

protein......... 4 calories/gram	carbohydrates ..... 4 calories/gram
fat .... 9 calories/gram	alcohol .... 7 calories/gram

WHAT FACTORS ARE INVOLVED IN THE EXPENDITURE OF CALORIES?

Surface area (related to height and weight): This is related to the amount of heat loss from the body. A lean individual has a greater surface area and therefore a greater energy requirement.

Age: The basal metabolic rate (BMR) is a measure of energy utilization at rest. BMR is related to growth and lean muscle mass. Infants and children are rapidly growing and thus have a higher BMR.

Sex: Females tend to have a lower BMR than males. This generalization arises from observations that women have a smaller percentage of lean muscle mass and that the female sex hormones may also affect metabolism.

Activity levels (exercise): Long-term effects are more important than the actual calories "burned" during exercise. Regular exercise can increase the BMR and therefore increase the overall calories expended. An exercise program designed to increase lean muscle mass will help increase the BMR (see the earlier section on endurance training).

CHAPTER 16: ANSWERS TO EVEN-NUMBERED REVIEW QUESTIONS

2.   a. Kidney, b. Liver, c. Intestine, d. Brain, e. Adipose tissue, f Liver.

4.   Hormones are molecules that function in intercellular communication, i.e., they regulate physiological responses throughout the body.

6.   The major recognized second messengers are (1) cAMP (generated from ATP by adenylate cyclase) stimulates changes in cellular activities by activating several protein kinases, (2) cGMP (generated from GTP by guanylate cyclase) activates protein kinase G, (3) diacylglycerol (DAG, a product of phospholipase C) activates protein kinase C, (4) inositol-1,4,5-triphosphate (IP3, a product of phospholipase C) that triggers the release of Ca^{2+} from the calcisome and (5) calcium ions which regulate the activities of numerous cellular activities when they bind to calcium-dependent regulatory proteins.

8.   Phorbol esters, found in croton oil, activate protein kinase C, an action that stimulates cell growth and division. However, unlike DAG, the molecule that they mimic, phorbol esters continue to activate protein kinase C for a prolonged time. This circumstance provides the affected cell with an advantage over unstimulated cells. Phorbol esters may transform a cell previously exposed to a carcinogenic initiating event into a cancerous cell whose unrestrained proliferation creates a tumor.

10.   In uncontrolled diabetes, large amounts of glucose are excreted in the urine. Excessive urine flow caused by the large amounts of water that are excreted along with the glucose dehydrates the body. Dehydration then usually triggers the thirst response.

12.   The hormones trigger increased protein synthesis in skeletal muscle among other metabolic changes.

14.   During the initial phase of a prolonged fast, blood glucose and insulin levels fall, and glucagon release is triggered. Glucagon acts to prevent hypoglycemia by promoting glycogenolysis and gluconeogenesis. The amino acids derived from muscle protein are a major source of the carbon skeleton substrates in gluconeogenesis.

16. HbA_{1c} formation is a consequence of nonenzymatic glycosylation of hemoglobin that occurs in the presence of high blood glucose levels. In the Maillard reaction the aldehyde group of glucose condenses with a free amino group in a protein to form a Schiff base. The Schiff base rearranges to form a stable ketoamine referred to as the Amadori product. The Amadori product subsequently destabilizes to form a reactive carbonyl-containing product that reacts with hemoglobin molecules to form an adduct such as HbA_{1c}.

CHAPTER 16: ANSWERS TO EVEN-NUMBERED THOUGHT QUESTIONS

2.   The hypothalamus regulates the function of the pituitary largely by secreting small amounts of specific releasing factors into a specialized capillary bed which directly connects the two structures. If the pituitary were to be transplanted to another part of the body, the concentration of the hypothalamic releasing factors in the blood that reaches the pituitary cells would be too low to affect their function.

4. Long-term fasting or low-calorie diets are interpreted by the body as starvation. The brain responds by lowering the body's BMR. One effect of this mechanism is that skeletal muscle mass is reduced because it is such a metabolically demanding tissue.

6. Increased mobilization of fatty acids provides an alternate energy source for muscle, thereby sparing glucose for the brain. In addition, glucagon stimulates gluconeogenesis, a pathway that utilizes amino acids derived from muscle.

8. To ensure proper control of metabolism, powerful hormones are synthesized in small quantities. Hormones also elicit responses in only specific target cells. They are metabolized quickly to ensure the precision of metabolic regulation.

10. The steroid molecule is covalently bound to the matrix in a chromatographic column. The extract suspected of containing the steroid binding protein is then passed through the column. Any proteins remaining on the column are eluted by changing the salt concentration of the eluting buffer. After their isolation and purification, such proteins can be examined specifically for binding activity to the steroid.

Nucleic Acids

THE CENTRAL DOGMA: REPLICATION, TRANSCRIPTION, AND TRANSLATION

The central dogma describes the flow of genetic information from DNA through RNA and eventually to proteins.

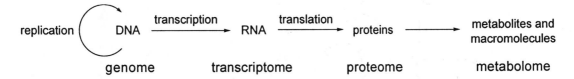

replication ⟶ DNA ⟶ transcription ⟶ RNA ⟶ translation ⟶ proteins ⟶ metabolites and macromolecules

genome   transcriptome   proteome   metabolome

DNA: AN ANTIPARALLEL DOUBLE HELIX

DNA contains all of an organism's genetic information,[1] and this information is encoded as a series of nitrogenous bases (purines and pyrimidines). Each base is connected to a 2'-deoxyribose molecule with a phosphate group at its 5' carbon. Remember (from Chapter 14) that this base-sugar-phosphate package is a nucleotide.[2]

DNA is a polymer of nucleotides linked together by phosphodiester bonds. (This is similar to the way that amino acids are linked by peptide bonds to form proteins, and sugars are linked by glycosidic bonds to form polysaccharides.) The phosphodiester bond is formed between the 5'-phosphate group and the 3'–OH on another nucleotide.

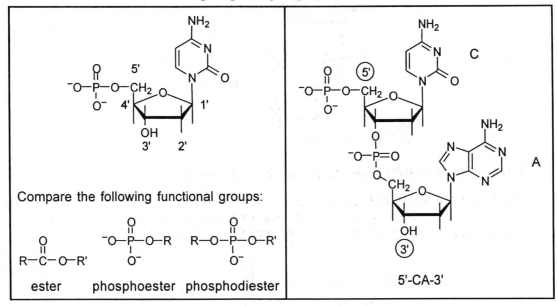

Compare the following functional groups:

ester   phosphoester   phosphodiester

5'-CA-3'

---

[1] Some viruses use RNA rather than DNA for their genetic information.

[2] Remember the difference between a nucleoside and a nucleotide: the -Side is the Sugar + base, and the -Tide is the Total Package, with the Phosphate, Too. Review the structures of purine and pyrimidine bases, nucleosides, and nucleotides in Chapter 14.

The pK_a of a phosphodiester is about 2, so at physiological pH (7) the phosphate is negatively charged. That makes nucleic acids polyanions (negatively charged). The negatively charged phosphates are usually associated with Mg^{2+} or cationic proteins (proteins with basic amino acids such as arginine or lysine).

The direction of a nucleic acid strand is given by the ends of the sugar which are not participating in a phosphodiester bond. The notation is 5'→3' or 3'→5'.

Nucleic acid sequences are written from the 5' to the 3' ends. For example, the sequence diagrammed below is 5'-CATG-3' or CATG.

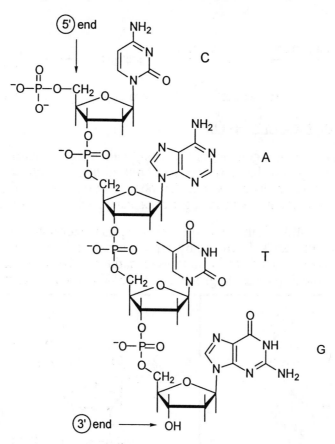

BASE-PAIRING ALLOWS DNA TO TRANSFER INFORMATION

The base from one strand can hydrogen bond to a base on the opposite strand. This is a *base pair,* or *bp*. Two hydrogen bonds can form between A and T (noted as A=T), whereas three hydrogen bonds form between C and G (noted as C≡G).

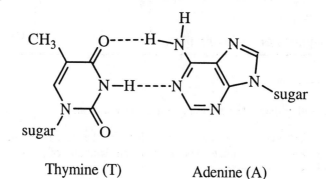

Thymine (T)   Adenine (A)

194

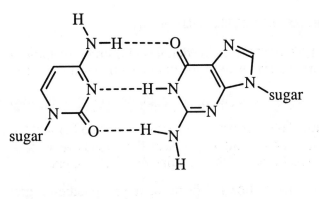

Cytosine (C)   Guanine (G)

Base pairing is specific (for example, A base-pairs with T but not with G or C). Other combinations are not possible due to:

1. **STERIC FACTORS:** A purine-purine base pair, like A=A, G=G, or A=G, would be bulky and cause the helix to be distorted. Likewise, a pyrimidine-pyrimidine base pair, such as C=C, T=T, or C=T, would also distort the double helix.

2. **HYDROGEN BONDING:** T cannot effectively hydrogen bond with G, and the same holds true for C and A. Try this out for yourself. Draw out the bases and see if you can get them to hydrogen bond to each other.

DNA STRANDS ARE COMPLEMENTARY

Because of this specificity of base pairing, the strands are complementary and can serve as templates for each other. This means that if the sequence of one strand is known, then the sequence of the other strand can be determined. Complementary copies can be made from a single strand. One strand can serve as the template for synthesis of the other strand.

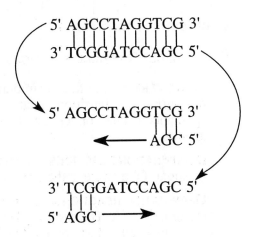

FEATURES OF THE WATSON-CRICK DNA DOUBLE HELIX (ALSO KNOWN AS B-DNA)

1. Two polynucleotide chains are twisted around each other to form a right-handed double helix with a diameter of 2.4 nm. Each helical turn occurs at 3.4 nm, with 10.4 bp per turn.

2. Base-pairing brings two nucleic acid strands together, and the chains are held together by hydrogen bonds between the bases.

3. Chains run in opposite directions or are antiparallel. That is, one strand is 5'→3' and the other strand is 3'→5'.

4. Nucleotide bases are on the inside of the helix, whereas the negatively-charged phosphate-sugar backbone are on the outside.

FORCES THAT STABILIZE THE DOUBLE HELIX INCLUDE:

HYDROPHOBIC INTERACTIONS: The double helix structure allows the hydrophobic purine and pyrimidine rings to reside in the interior and away from water.

HYDROGEN BONDING: Specific hydrogen bonds can form between the bases.

BASE STACKING: The base pairs are parallel to each other and therefore allow van der Waal contacts. For an individual base, the force is *very* weak. But when a DNA molecule has ≈10,000 bases, the cumulative effect is great.

ELECTROSTATIC INTERACTIONS: The negative phosphodiester groups tend to repulse each other. This is partly why the helix is twisted. Its charge needs to be stabilized or shielded by cations such as Mg^{2+}, polyamines, and histones.

DNA Structure: The Nature of Mutation

MUTATIONS are permanent changes in the base sequence of DNA molecules.

POINT MUTATIONS involve a single base pair. Types of point mutations:

TRANSITION MUTATION: a pyrimidine substitutes for another pyrimidine, or a purine substitutes for another purine.

TRANSVERSION MUTATION: a pyrimidine substitutes for a purine, or vice versa.

What causes mutations?

SPONTANEOUS CHEMICAL REACTIONS

TAUTOMERIC SHIFTS that occur during DNA replication can cause base mispairing and transition mutation. (Remember that tautomeric shifts are the spontaneous interconversion of amino and imino groups or of keto and enol groups.)

DEPURINATION REACTIONS: spontaneous hydrolysis that cleaves a purine from its sugar. One cause is the protonation of guanine's N-3 and N-7.

DEAMINATION REACTIONS. *Example*: C can change to U via deamination and a tautomeric shift. The resulting U would base-pair with A (rather than the G that was supposed to base-pair with C).

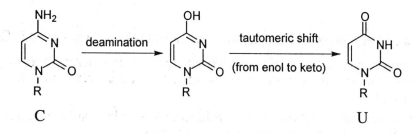

IONIZING RADIATION SUCH AS UV, X-RAYS, AND γ-RAYS

Free-radical mechanisms cause strand breaks, DNA-protein crosslinking, ring openings, base modifications (thymine glycol, 5-hydroxymethyl uracil, 8-hydroxyguanine), formation of dimers between two pyrimidines (thymine dimers are caused by UV radiation).

XENOBIOTICS:

BASE ANALOGUES have structures that are similar to normal nucleotide bases, and can be inadvertently incorporated into DNA.

ALKYLATING AGENTS add carbon-containing alkyl groups that may promote tautomer formation.

NONALKYLATING AGENTS modify DNA structure in other ways, and typically prevents base pairing. Examples; HNO_2 deaminates bases, and benzo[a]pyrene is metabolized to a highly reactive intermediate that forms adducts to bases.

INTERCALATING AGENTS can insert themselves between stacked base pairs, distorting the DNA double helix and causing deletion or insertion of base pairs (resulting in frame-shift mutations).

DNA Structure: From Mendel's Garden to Watson and Crick

HISTORICAL DEVELOPMENTS THAT LED TO THE DISCOVERY OF DNA STRUCTURE:

1865	The scientific revolution that eventually necessitated the determination of DNA structure began when Gregor Mendel discovered the basic rules of inheritance.
1869	Friedrich Miescher discovered "nuclein," later renamed nucleic acid.
1882-1897	The chemical composition of DNA was determined largely by Albrecht Kossel.
1928	Fred Griffith proposed the concept of transmission of genetic information between bacterial cells.
1944	Avery and McCarty demonstrated that the digestion of DNA by deoxyribonuclease inactivated the transforming agent, concluding that genetic information was carried by DNA.
1952	By using T2 bacteriophage, Hershey and Chase demonstrated the separate functions of viral nucleic acid and protein. These experiments reconfirmed DNA as the genetic material.

INFORMATION USED BY WATSON AND CRICK TO DETERMINE DNA STRUCTURE

By the early part of the decade, the chemical structures and dimensions of the nucleotides had been elucidated, and it was known that adenine:thymine and guanine:cytosine existed as 1:1 ratios in DNA. X-ray diffraction studies performed by Rosalind Franklin indicated that DNA was a symmetrical molecule, in all likelihood a helix; the diameter and pitch of the helix was estimated by Wilkins and Stokes. Also, Linus Pauling had recently shown that protein could exist as a helix.

DNA Structure: Variations on a Theme

B-DNA: Base pairs are at right angles; right-handed helix with 10.4 bp per turn; each helical turn occurs at 3.4 nm; and the diameter is 2.4 nm. B-DNA occurs as the sodium salt of DNA under highly humid conditions.

A-DNA: Base pairs tilt 20° away from the horizontal; right-handed helix with 11 bp per turn; each helical turn occurs at 2.5 nm; and the diameter is 2.6 nm. A-DNA occurs when DNA becomes partially dehydrated, as when extracted with solvents such as ethanol.

Z-DNA: Base pairs are in a zigzag conformation; left-handed helix with 12 bp per turn; each helical turn occurs at 4.5 nm; and the diameter is 1.8 nm. Base-pairs with alternating pyrimidine-purine bases are most likely to be in Z-DNA configuration.

HIGHER-ORDER STRUCTURES:

CRUCIFORMS - crosslike structures; likely when a DNA sequence contains a palindrome (inverted repeats)

H-DNA: Triple helix, formed in certain circumstances between a long DNA segment consisting of a polypurine strand hydrogen-bonded to a polypyrimidine strand. H-DNA formation depends on non-conventional base-pairing that occurs without disrupting the Watson-Crick base pairs.

SUPERCOILING: packages a large DNA molecule into a compact form.

DNA Supercoiling

DNA supercoiling facilitates several biological processes such as DNA replication and transcription.

NEGATIVE SUPERCOILING:

Underwound DNA twists to the right to relieve strain.

DNA winds around itself to from an interwound supercoil.

POSITIVE SUPERCOILING:

Overwound DNA twists to the left to relieve strain.

DNA winds around a protein core to form a toroidal[3] supercoil.

Chromosomes and Chromatin

RELATED TERMS:

CENTROMERES - structures that attach chromosomes to the mitotic spindle during mitosis and meiosis

TELOMERES - structures at the end of chromosomes that buffer the loss of critical coding sequences after a round of DNA replication

TRANSPOSITION - transposons (transposable DNA sequences) excise themselves and insert at another site (or, involve an RNA transposon or retrotransposon)

PROKARYOTIC CHROMOSOME IS CIRCULAR DNA...

...that's extensively looped and coiled. (*Example*: An *E. coli* nucleoid is supercoiled DNA complexed with a protein core.)

---

[3] A toroid is the shape of a donut.

EUKARYOTES: HUMANS POSSESS 23 PAIRS OF CHROMOSOMES

CHROMOSOME: one linear DNA molecule complexed with histones; structural units are the nucleosomes

HISTONES: a group of small basic proteins found in all eukaryotes; positively-charged Arg in histones interact with negatively-charged phosphates in DNA

NUCLEOSOME: eight histone molecules with about 140 base pairs of DNA wrapped around the histone by one-and-three-quarters turns. An additional sixty base pairs connect adjacent nucleosomes.

Nucleosomes are coiled into **30-NM FIBERS**, which are further coiled to form **200-NM FILAMENTS**.

CHROMATIN - partially decondensed form of chromosomes; occurs when cell is not dividing

ORGANELLE DNA - IN MITOCHONDRIA AND CHLOROPLASTS

Genome Structure

A genome is a complete set of genetic information, encoded in the nucleotide base sequence of its DNA.

PROKARYOTIC VS. EUKARYOTIC GENOMES

Feature:	Prokaryotic	Eukarotic
Genome size	relatively small, fewer genes	relatively large (but size doesn't imply complexity of an organism)
Coding capacity	genes are compact and continuous; little to no noncoding DNA	majority of DNA is noncoding
Gene expression	operon - set of functionally related genes that are regulated as a unit	introns interspersed between exons; introns = noncoding sequences; exons = expressed sequences.
Other unique features	plasmids - additional DNA, typically circular, that codes for biomolecules that provide a growth or survival advantage	contains repetitive sequences (see below)

TYPES OF REPETITIVE SEQUENCES IN EUKARYOTIC GENOMES

TANDEM REPEATS - multiple copies arranged next to each other

INTERSPERSED GENOME-WIDE REPEATS - repetitive sequences scattered around the genome; most result from transposition

RNA: STRUCTURE, FUNCTION, AND SYNTHESIS (TRANSCRIPTION)

Differences between RNA and DNA:

1. The sugar of RNA is ribose, so there is a 2'–OH group on the sugar.
2. Uracil replaces thymine. So, the base pairs in RNA are: A=U and C≡G.
3. RNA is a single strand (not a double helix).

Types of RNA:

Of the total RNA of the cell, ≈80% is rRNA, ≈15% is tRNA, ≈5% is mRNA, with small amounts of heterogeneous and small nuclear RNA.

1. Ribosomal (rRNA) - an integral part of ribosomes.

2. Transfer (tRNA) - carries individual amino acids to ribosomes, where they are properly aligned and assembled into proteins.

3. Messenger (mRNA) - carries the genetic information for protein synthesis from the nucleus to the ribosomes. mRNA molecules are copies of DNA "genes," and contain the genetic information needed to synthesize specific polypeptides.

4. Heterogeneous RNA (hnRNA) - precursors to mRNA

5. Small nuclear RNA (snRNA) - involved in RNA splicing and other processing

Transfer RNA (tRNA): Adapter Molecules in Protein Synthesis

tRNA are single-stranded polynucleotide chains about 75 nucleotides long that convert a nucleotide sequence into a specific amino acid. tRNA contains modified bases that help to stabilize and protect tRNA from degradation. These modifications can also occur in rRNA.

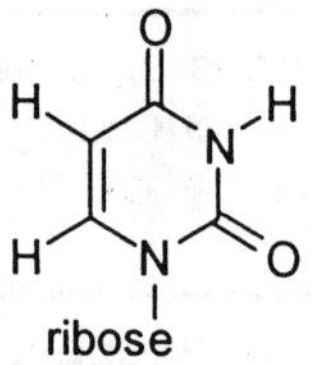 uridine	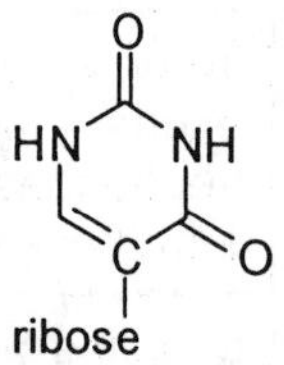 pseudouridine Ψ	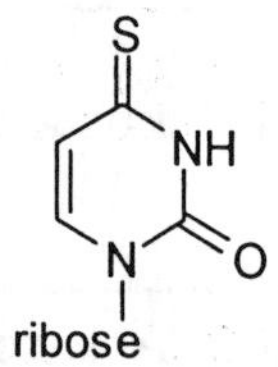 4-thiouridine	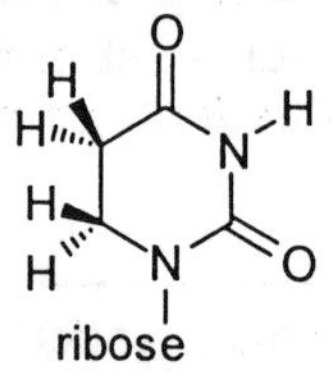 dihydrouridine	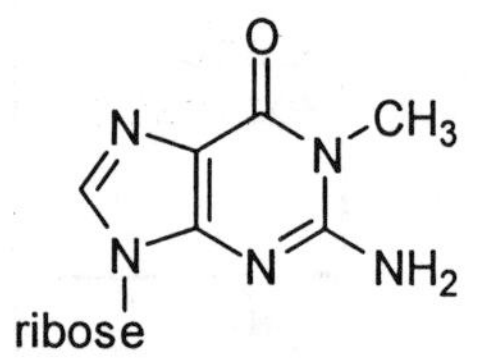 1-methyl guanosine
(shown for comparison)	Attachment to ribose is via a carbon atom.	S replaces a carbonyl O.		

THE SHAPE OF tRNA RESEMBLES A WARPED CLOVERLEAF WITH THE FOLLOWING FEATURES:

3'-TERMINUS - bonds to a specific amino acid

ANTICODON LOOP - contains a sequence of 3 base pairs that complement the DNA triplet code for a specific amino acid.

D LOOP - contains dihydrouridine

TΨC LOOP - contains thymine, pseudouridine, and cytosine

VARIABLE LOOP - length varies between 4-5 to 20.

Ribosomal RNA (rRNA) is an integral component of ribosomes

Ribosomes are cytoplasmic structures that synthesize proteins and consist of 60-65% rRNA, with the remainder being protein. Ribosomes consist of two subunits of unequal size, with several different kinds of rRNA and protein in each subunit.

Messenger RNA (mRNA) carries DNA's genetic info for protein synthesis.

specifies the order of amino acids in a protein.

cistron - a DNA sequence that contains coding info for a polypeptide plus signals needed for ribosome function

PROKARYOTIC VS. EUKARYOTIC MRNA:

POLYCISTRONIC (prokaryotes) vs. **MONOCISTRONIC** (eukaryotes): contain information for many (poly) or one (mono) polypeptide chains

Prokaryotic mRNA isn't process further, and can begin working right away, as-is. Eukaryotic mRNA is extensively modified. Eukaryotic mRNA modifications include:

- Capping of the 5' end with a 7-methylguanosine
- Splicing - removing introns
- Attaching a polyadenylate ("poly A tail") to the 3' end

Heterogeneous RNA (hnRNA) and Small Nuclear RNA (snRNA)

HNRNA - primary transcripts of DNA, precursors to mRNA (hnRNA is spliced and modified to form mRNA)

SNRNA - involved in splicing and other RNA processing; complexed with several proteins to form small nuclear ribonucleoprotein particles (snRNP or snurps)

Splicing is the enzymatic removal of introns from the primary transcripts.

VIRUSES: OBLIGATE, INTRACELLULAR PARASITES OR MOBILE GENETIC ELEMENTS?

Each virus contains a piece of nucleic acid within a protective coat. When a virus infects a host cell, its nucleic acid hijacks the cell's nucleic acid and protein-synthesizing machinery, making complete new viral particles that are released form the host cell (often rupturing the cell in the process). Or, the viral nucleic acid may insert itself into a host chromosome, resulting in cell transformation.

WHAT PROPERTIES OF VIRUSES MAKE THEM USEFUL RESEARCH TOOLS?

Driven in large part by the role viruses play in numerous diseases, viral research has been of tremendous benefit to biochemistry. Because a virus essentially subverts normal cell function to produce more viruses, a viral infection can provide unique insight into cellular metabolism. Several eukaryotic genetic mechanisms have been elucidated with the aid of viruses and/or viral enzymes. Viral research has also provided substantial information concerning genome structure and carcinogenesis. Viruses have also been invaluable in the development of recombinant DNA technology.

The Structure of Viruses

VIRIONS - complete viral particles = nucleic acid + capsid (+ membrane envelope in more complex viruses)

CAPSID - a protein coat made of interlocking CAPSOMERES (protein molecules)

NUCLEOCAPSID - nucleic acid + capsid

TYPES OF VIRAL GENOMES:

Most common: double-stranded DNA (dsDNA) or single-stranded RNA (ssRNA: positive-sense vs. negative-sense)

Also observed: single-stranded DNA (ssDNA), double-stranded RNA (dsRNA)

AFTER MANY YEARS OF RESEARCH AND EXPENDING BILLIONS OF DOLLARS, WHY IS AIDS STILL CONSIDERED AN INCURABLE DISEASE?

HIV is a retrovirus that is believed to cause AIDS. Retroviruses are a class of RNA viruses which possess a reverse transcriptase activity that converts their RNA genome to a DNA molecule. This DNA is then inserted into the host cell genome, causing a permanent infection.

Because the viral genome mutates frequently (i.e., its surface antigens become altered), a vaccine for the HIV virus has been difficult to develop. In the case of HIV, mutations occur because the reverse transcriptase doesn't have any proofreading capabilities. So, now and then it makes mistakes that are related to the surface antigens, which are continuously altered.

Similarly, that's why we need to get a flu shot every year. Flu viruses are very active, and mutate constantly because of continuous replication. Their enzymes make mistakes that are beneficial for the survival of the virus.

CHAPTER 17: ANSWERS TO EVEN-NUMBERED REVIEW QUESTIONS

2. B-DNA, the right handed double helical structure discovered by Watson and Crick, in which there are 10.4 base pairs per helical turn (3.4 nm; diameter = 2.4 nm), occurs under humid conditions. A-DNA, a partially dehydrated molecule, possesses 11 base pairs per helical turn (2.5 nm; diameter 2.6 nm). The Z form of DNA is twisted into a left handed spiral with 12 base pairs per helical turn. Each helical turn occurs in 4.5 nm with a diameter of 1.8 nm.

4. Biological processes that are facilitated by supercoiling include packaging of DNA into compact forms (i.e., chromosomes), DNA replication and transcription.

6. Nucleosomes are the structural units of chromatin. Each nucleosome consists of a left-handed supercoiled DNA segment wound around eight histone molecules.

8. The three major types of RNA are ribosomal RNA (a component of ribosomes), transfer RNA (each molecule transports a specific amino acid to the ribosome for assembly into proteins), and messenger RNA (each molecule specifies the sequence of amino acids in a polypeptide).

10. There are approximately 6 million base pairs in a single human cell. Assuming that there are 10^{14} body cells, the total length of the DNA in the human body is approximately 2×10^{11} km. This estimate is about 1000 times greater than the distance from the earth to the sun.

12. Guanine-cytosine pairs contain three hydrogen bonds while adenine-thymine contain only two. The more hydrogen bonds holding the DNA strands together, the higher the melting point will be.

14. The hierarchy from smallest to largest structure is (d) nucleotide base pair, (c) nucleosome, (b) gene and (a) chromosome

16. a. telomere – structure at the ends of chromosomes that buffer the loss of critical coding sequences after a round of DNA replication.

    b. minisatellite – tandemly repeated DNA sequence of about 25 bp with total lengths between 10^2 and 10^5 bp.

    c. DNA profile – consists of the unique pattern and number of repeats each in STR sequences used to identify individuals.

    d. intron – a noncoding intervening sequence in a split or interrupted gene missing in the final RNA product

    e. exon – a region in a split or interrupted gene that codes for RNA and ends up in the final product (e.g., mRNA).

    f. retrotransposon – a transposition mechanism that involves an RNA transcript; an RNA transposon

    g. DNA denaturation – the separation of the DNA double helix into single strands.

    h. chain-terminating method – a DNA sequencing method developed by Frederick Sanger that uses 2',3'-dideoxy nucleotide derivatives that inhibit chain termination

    i. PCR – polymerase chain reaction; a method for rapidly producing large numbers of copies of DNA sequences.

    j. splicing – the excision of introns during mRNA processing.

CHAPTER 17: ANSWERS TO EVEN-NUMBERED THOUGHT QUESTIONS

2.   The major and minor grooves of DNA arise because the glycosidic bonds in the two hydrogen bonded strands are not exactly opposite to each other.

4.   RNA can coil back on itself to form complex three-dimensional structures.

6.   The electron withdrawing effect of the bromine increases the likelihood of enol formation of uracil. This enol mimics the hydrogen bonding pattern of cytosine. Therefore, this base can be paired with guanine.

8.   One of the principal reasons for the problems in the development of an AIDS vaccine is that the HIV genome mutates frequently. Consequently, the surface antigens of HIV also often change. Note that the antigens in a vaccine stimulate the immune system to produce antibodies that will bind specifically to antigens on the surface of disease-causing organisms.

10.   The histones act to shield the DNA from the action of nucleases.

12.   Each base in the nucleotides of DNA and RNA has a specific shape that has unique information content. The enormous number of possible sequences of the nucleotides make possible a very large coding capacity that living organisms use to specify the molecular structure of all of their biomolecules.

Genetic Information

REPLICATION, REPAIR, AND RECOMBINATION

DNA Replication: DNA Polymerase Makes Copies of DNA

GENERAL FEATURES OF DNA REPLICATION:

1. The two strands of DNA are separated by unwinding.

2. Semiconservative replication: Each strand serves as a template for the synthesis of a complementary strand, so each new DNA molecule has one old strand and one new strand. (This was proven in the Meselson-Stahl experiment.) An existing DNA template is essential to provide the proper sequence of bases.

3. Deoxyribonucleotides are added to the 3'-hydroxyl of a pre-existing nucleic acid chain (either DNA or RNA) by forming new phosphodiester bonds.

4. Deoxyribonucleotide triphosphates (dNTPs) such as dATP, dGTP, dCTP, and dTTP, and Mg^{2+} are the substrates for DNA polymerase.

$$(DNA)_n + dNTP \rightarrow (DNA)_{n+1} + PP_i$$

5. Elongation proceeds only in the $5' \rightarrow 3'$ direction. Because of this, one strand, the **leading strand**, can be synthesized continuously, while the other strand, the **lagging strand**, is only synthesized discontinuously, as fragments.

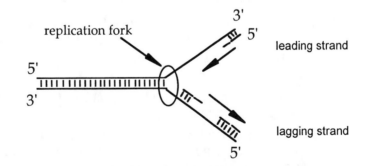

The synthesis of the new leading strand proceeds toward the replication fork. As the replication fork opens up, the leading strand is continuously synthesized.

The synthesis of the lagging strand is a bit more complex. Because DNA synthesis can **only** proceed in the $5' \rightarrow 3'$ direction, a new RNA primer needs to be synthesized as the replication forks open up. Therefore, the lagging strand is synthesized in short segments or fragments called Okazaki fragments.

DNA SYNTHESIS IN PROKARYOTES (*E. COLI*)

1. ### DNA UNWINDING UTILIZES THREE TYPES OF PROTEINS:

   Topoisomerases work ahead of the replication fork to relieve torque. (See supercoiling control, # 5, below.)

   Helicases unwind the DNA at the fork (and require ATP).

   SSB (single-stranded binding protein) stabilizes and protects single-stranded DNA segments.

   Prokaryotic replication begins at oriC (initiation site) and proceeds in two directions, forming a replication eye. Replicon - a DNA molecule or segment that contains an initiation site and regulatory sequences

2. ### PRIMERS - SHORT RNA SEGMENTS NEEDED TO INITIATE REPLICATION - ARE SYNTHESIZED BY PRIMASE

   An RNA primer provides a 3'-OH group, to which DNA polymerase can add its first nucleotide.

   Primosome: a multienzyme complex containing primase (an RNA polymerase) and several auxiliary proteins.

   The leading strand only needs one primer per replication fork, but the lagging strand needs a primer for each Okazaki fragment.

3. ### DNA SYNTHESIS IS CATALYZED BY DNA POLYMERASES

   Replisome: the prokaryotic DNA replicating machine: DNA unwinding proteins, the primosome, and two copies of the DNA polymerase III.

   DNA polymerases are large multienzyme complexes that catalyze the formation of phosphodiester bonds between dNTPs in the $5' \rightarrow 3'$ direction.

   DNA polymerase III (pol III) (in prokaryotes) has at least 10 subunits.

   - **CORE POLYMERASE** - 3 subunits: α, ϵ, θ: α subunit forms the phosphodiester bonds ($5' \rightarrow 3'$ polymerase); ϵ subunit has $3' \rightarrow 5'$ exonuclease activity (see below); θ subunit function is unknown.

   - **SUBUNIT τ** - forms a dimer of 2 cores

   - **β-PROTEIN** - sliding clamp protein - (2 subunits) - forms a ring around the template DNA strand;

   - **γ COMPLEX** - (5 subunits) - transfers β-protein to the core polymerase; promotes processivity (keeps the polymerase with the template DNA during replication) because the β-protein acts like a tether between the polymerase and the template DNA

   The RNA primers used in DNA replication need to be removed and replaced by DNA. This is where the $5' \rightarrow 3'$ exonuclease activity of DNA polymerase I (pol I) comes into play. Pol I degrades the RNA primers ahead of it as the enzyme synthesizes DNA. RNA primer sequences are replaced with DNA sequences.

OTHER DNA POLYMERASES:

DNA POLYMERASE I (pol I, Kornberg enzyme) - DNA repair enzyme; also removes of RNA primer, especially on the lagging strand

DNA POLYMERASE II (pol II) - similar to pol I, but poorly understood

EXONUCLEASE ACTIVITY - NUCLEOTIDE REMOVAL FROM AN END OF A POLYNUCLEOTIDE STRAND

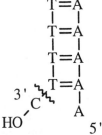

$3' \to 5'$ EXONUCLEASE ACTIVITY functions as a "proofreader." If pol III adds an incorrect base to the newly synthesized strand, the $3' \to 5'$ exonuclease activity (of pol I or pol III) will remove the mistake and let the polymerase try again.

Some DNA polymerases can also hydrolyze nucleotides from the 3' end, but only when there is a free 3'-hydroxyl group and when the nucleotide is not part of the double helix (not base paired). All three DNA polymerases have $3' \to 5'$ exonuclease activity.

$5' \to 3'$ EXONUCLEASE ACTIVITY functions to remove primers or to repair damaged DNA. The nucleotide to be removed must be part of the double helix; the polymerase removes the nucleotides in front of it. Pol I has this activity.

Replication ends when replication forks meet at the ter region (termination site).

4. JOINING OF DNA FRAGMENTS BY DNA LIGASE, which catalyzes the formation of a new phosphodiester bond. Discontinuous DNA synthesis of the lagging strand requires DNA ligase to join the newly synthesized fragments.

5. SUPERCOILING CONTROL: DNA topoisomerases prevent the DNA strands from tangling, which would prevent further unwinding of the double helix.

   DNA topoisomerases change the supercoiling of DNA by breaking one or both strands, passing the DNA through the break and rejoining the strands. This functions to relieve torque.

   Controlled supercoiling can facilitate DNA unzipping.

   Type I topoisomerases: make transient single-strand breaks in DNA

   Type II topoisomerases: make transient double-strand breaks

   DNA gyrase: a prokaryotic type II topoisomerase that helps to separate the replication products and to create the negative supercoils needed for genome packaging

DIFFERENCES BETWEEN PROKARYOTIC AND EUKARYOTIC DNA REPLICATION ...

... appear to be related to the size and complexity of eukaryotic genomes.

TIMING OF REPLICATION: Eukaryotic cells only replicate during a specific period of time during their cell cycle (the S phase). (The eukaryotic cell cycle includes times of rest - G_0, G_1, G_2 - in addition to the S phase and the cell division phase.) In contrast, prokaryotes replicate throughout most of their cell cycle.

REPLICATION RATE: Because of the complex structure of chromatin, DNA replication is significantly slower in eukaryotes (about 50 nucleotides/second per replication fork) than in prokaryotes (about 500).

MULTIPLE REPLICONS are used by eukaryotes to compress the replication of their large genomes into short time periods. Instead of replisomes, eukaryotes have replication factories – immobilized sites that contain a large number of replication complexes. DNA is threaded through these complexes as it's synthesized.

OKAZAKI FRAGMENTS: Eukaryotic Okazaki fragments are significantly shorter (100-200 nucleotides) than prokaryotic fragments (1000-2000 nucleotides).

EUKARYOTIC REPLICATION ENZYMES

- Five types of eukaryotic DNA polymerase: α, β, δ, ϵ, γ
- DNA polymerase α: initiates synthesis of both leading and lagging strands
- DNA polymerase δ: two complexes: one synthesizes the leading strand, and one synthesizes the lagging strand; binds to PCNA, a sliding clamp protein that acts like β-protein in E. coli
- Replication protein A (RPA) (acts like SSB): keeps DNA strands separated during DNA synthesis
- FEN1 (MF1) - removes RNA primer from each Okazaki fragment; activity is associated with the δ complex.
- DNA ligase joins Okazaki fragments
- DNA polymerase β: involved in DNA repair
- DNA polymerase ϵ: $3' \rightarrow 5'$ exonuclease activity; other functions unknown
- DNA polymerase γ: catalyzes mitochondrial genome replication
- Eukaryotic type II topoisomerases catalyze only the removal of superhelical tension (refer to the description of prokaryotic DNA gyrase, above).

DNA Repair Mechanisms[1]

DNA ligase can repair breaks in the phosphodiester linkages.

PHOTOACTIVATION REPAIR (also called light-induced repair) of pyrimidine dimers: DNA photolyase[2] has flavin and pterin chromophores that use energy from visible light to cleave pyrimidine dimers, restoring them to two separate pyrimidines and leaving the phosphodiester bonds intact.

EXCISION REPAIR: Incorrect bases are removed (excised) and replaced with the correct ones. Excision repair involves a series of enzymes.

A repair endonuclease (also, excision nuclease or excinuclease) detects a distorted DNA segment, then cuts the damaged DNA and removes a single-stranded

---

[1] (At this point, it might be helpful to review the types of DNA mutations that were discussed in Chapter 17. Remember what causes pyrimidine dimers and which bonds needs to be cleaved to separate them? ☺)

[2] DNA photolyase is not present in humans. ☹

sequence about 12 nucleotides[3] in length. The enzyme pol I replaces the nucleotides in the gap left by the excised segment, and DNA ligase seals the break in the phosphodiesterase backbone.

RECOMBINATIONAL REPAIR: Post-replication repair of DNA. Damaged DNA interrupts replication; the replication complex detaches from the DNA and re-initiates after the damaged site. This results in a gap in the daughter strand. This gap is repaired by an exchange of the corresponding segment of the homologous DNA (this process is called recombination). After recombination, DNA polymerase and DNA ligase complete the repair process.

DNA Recombination *produces new combinations of genes (& fragments): variations that make evolution possible*

DNA recombination is the rearrangement of DNA sequences by exchanging segments from different molecules

GENERAL RECOMBINATION requires the precise pairing of homologous DNA molecules.

SITE-SPECIFIC RECOMBINATION requires only short regions of DNA homology; depend more on protein-DNA interactions than on sequence homology

TRANSPOSITION - a variation of site-specific recombination - transposable elements (certain DNA sequences) are moved from one chromosome or chromosomal region to another; differs from site-specific recombination in that a specific protein-DNA interaction occurs on only one of the 2 recombining sequences. The recombination of the 2nd DNA sequence is nonspecific.

GENERAL RECOMBINATION

1. Pairing of two homologous DNA molecules
2. Nicking: Two of the DNA strands (one in each molecule) are cleaved.
3. Crossover: The two strand segments cross over, forming a Holliday intermediate
4. Sealing nicks: DNA ligase seals the cut ends
5. Branch migration caused by base pairing exchange leads to the transfer of a segment of DNA from one homologue to the other.
6. A second series of DNA strand cuts occurs.
7. DNA polymerase fills any gaps, and DNA ligase seals the cut strands.

FORMS OF INTERMICROBIAL DNA TRANSFER IN BACTERIA:

1. **TRANSFORMATION** - naked DNA fragments enter through an opening in the cell wall and are introduced into the bacterial genome

2. **TRANSDUCTION** - bacteriophage inadvertently carry bacterial DNA to a recipient cell; after recombination, the cell uses the transduced DNA

---

[3] 27-29 in eukaryotes

3. CONJUGATION - an unconventional sexual mating: A donor cell synthesizes a sex pilus (via a specialized plasmid) that attaches to the surface of the recipient cell. The pilus transfers a fragment of the donor's DNA, which can undergo recombination or exist in plasmid form.

SITE-SPECIFIC RECOMBINATION

Example: Integration of bacteriophage λ into the *E. coli* chromosome (See Figure 18.15, p. 626 of your text.)

TRANSPOSITION

TRANSPOSONS - transposable elements (jumping genes) that can jump between bacterial chromosomes, plasmids, and viral genomes

IS ELEMENTS - insertion elements - bacterial transposons that consist only of a gene that codes for a transposase (transposition enzyme) flanked by inverted repeats (short palindromes)

COMPOSITE TRANSPOSONS - bacterial transposons that contain additional genes, several of which may code for antibiotic resistance (!)

1. REPLICATIVE TRANSPOSITION - transposon inserts a replicated copy, stays in its original site

2. NONREPLICATIVE TRANSPOSITION - transposon is cut out of its original (donor) site and inserted into the target site; the donor site must be repaired

EUKARYOTIC TRANSPOSONS: many contain LTR (long terminal repeats or delta repeats); many mechanisms involve an RNA intermediate and resemble the replicative phase of a retrovirus

TRANSCRIPTION: RNA SYNTHESIS

CATALYZED BY RNA POLYMERASE: $NTP + (NMP)_n \rightarrow (NMP)_{n+1} + PP_i$.

NONTEMPLATE STRAND = plus (+) strand; also called the coding strand because it has the same base sequence as the RNA transcription product (except U substitutes for T)

TEMPLATE STRAND = minus (–) strand

The direction of the gene = the direction of the coding strand. So, polymerization proceeds from the 5' end to the 3' end of both the coding strand and the gene.

Transcription in Prokaryotes

RNA POLYMERASE IN E. COLI: core enzyme catalyzes RNA synthesis; sigma factor binds transiently to the core enzyme, and allows it to bind both the correct template strand and the proper site to initiate transcription

Pribnow box - region 10 nucleotides before the transcription initiation site

STAGES OF TRANSCRIPTION IN *E. COLI*:

1. Initiation: RNA polymerase binds to a promoter (a specific DNA sequence)

   short DNA segment near the Pribnow box unwinds

   first nucleoside triphosphate binds to the RNA polymerase complex, beginning transcription; attacks second NTP to form the first phosphodiester bond

   After transcribed sequence is about 10 nucleotides long, conformation of RNA polymerase complex changes; σ factor is detaches, RNA polymerase affinity for the promoter site decreases; initiation phase ends

2. Elongation phase: core RNA polymerase converts to an active transcription complex, binds several accessory proteins

   DNA unwinds ahead of the transcription bubble (topoisomerases resolve positive and negative supercoiling ahead of and behind the bubble)

   Elongation continues until a termination sequence is reached

3. Termination: Termination sequences contain palindromes, and their RNA transcripts form a stable hairpin turn.

PRODUCTS OF TRANSCRIPTION:

mRNA: used immediately for protein synthesis

tRNA and rRNA require post-transcriptional processing

Transcription in Eukaryotes… is significantly more complex than for prokaryotes.

UNIQUE FEATURES OF EUKARYOTIC TRANSCRIPTION:

1. RNA polymerase activity: requires three nuclear RNA polymerases:

   RNA polymerase I: transcribes large rRNAs

   RNA polymerase II: transcribes precursors of mRNA and most snRNAs

   RNA polymerase III: transcribes precursors of tRNAs and 5S rRNA

   Eukaryotic RNA polymerases can't initiate transcription. Various transcription factors must be bound at the promoter before transcription can begin.

2. **PROMOTERS:** larger, more complicated, and more variable than prokaryotic promoters. Many promoters for RNA polymerase II contains consensus sequences (TATA box) about 25-30 bp upstream from the initiation site.

   CAAT BOX, GC BOX - examples of sequences upstream that bind transcription factors and affect the frequency of transcription initiation

   ENHANCERS - regulatory sequences that may be thousands of bp away from the gene, but affect activity of promoters

3. **POSTTRANSCRIPTIONAL PROCESSING:** Eukaryotic mRNAs are extensively processed (unlike prokaryotic mRNAs, which typically have little to no post-transcriptional processing).

WHY MODIFY MRNA? To help mRNA transport out of the nucleus and to help stabilize the mRNA molecule.

EUKARYOTIC MRNA MODIFICATIONS INCLUDE:

CAPPING of the 5' phosphate end with a 7-methylguanosine protects the 5' end from exonucleases and promotes mRNA translation. Also, the first two nucleotides of the transcript are methylated at the 2'–OH.

ATTACHING A POLYADENYLATE ("poly A tail" with 100-250 As) to the 3' end protects from the action of 3',5'-exonucleases and promotes mRNA export to the cytoplasm

SPLICING - removing introns (DNA sequences that intervene and occur within a particular gene); Each intron is excised as a lariat (a configuration that resembles a loop); then the exons are joined.

Exons - coding sequences (DNA sequences that designate the amino acid sequence)

Not all of the DNA sequences in eukaryotes code for amino acids. Some DNA sequences serve regulatory or structural roles. So, the removal of introns is necessary to synthesize an mRNA that will be translated into a continuous (and correct) amino acid sequence

Spliceosome - where splicing happens; multi-component structure with several snRNAs, several proteins

Ribozyme - catalytic RNA that exhibits self-splicing; found in several organisms.

GENE EXPRESSION

INDUCIBLE GENES are expressed (turned on) only under certain circumstances.

CONSTITUTIVE OR HOUSEKEEPING GENES - transcribed routinely because they're needed for cell function

OPERONS - groups of linked structural and regulatory genes that control inducible genes

Most gene expression mechanisms involve DNA-protein binding. Structural similarities in many DNA regulatory proteins:

Twofold axis of symmetry; often form dimers

Supersecondary structures that interact with DNA: helix-turn-helix; helix-loop-helix; leucine zipper; zinc finger

DEREPRESSION: activating a gene by inhibiting a repressor (It's like taking your foot off the brake in a stopped car that's pointed downhill. It might not be the same as hitting the gas, but the car still moves forward!)

Gene Expression in Prokaryotes

REGULATION OF LACTOSE[4] METABOLISM IN *E. COLI* CELLS BY THE LAC OPERON

SUMMARY: When lactose is present, β-galactosidase converts a few molecules to allolactose, which turns the lac operon on. The lac operon codes for lactose metabolism enzymes until the lactose supply is consumed. Then, the repressor protein can once again bind to the operator site, and the lac operon turns off.

lac operon	structural genes Z, Y, and A that code for lactose enzymes; plus a control element - a promoter site that contains the CAP site and overlaps the operator site
CAP site	where the CAP protein binds
operator site	DNA sequence that binds a repressor protein and helps to regulate adjacent genes
repressor gene i	codes for the lac repressor protein when lactose is absent; directly adjacent to lac operon
lac repressor	protein that binds to the operator and prevents binding of RNA polymerase to the promoter (that is, it turns the lac operon off)
allolactose	β-1,6-isomer of lactose, and an inducer. When allolactose binds to the lac repressor, its conformation changes and it dissociates from the operator. When the operator is free of the repressor, the lac operon is turned on, and transcription of the structural genes can begin. In the absence of inducer, the lac operon remains off.

Glucose is a preferred carbon and energy source for *E. coli*. An organism that has both glucose and lactose will use glucose first. Only after the glucose is gone will the lac operon enzymes be synthesized. This is how it happens:

When glucose is depleted (and the cell needs energy), cAMP levels in the cell rise.[5] cAMP binds to CAP (catabolite gene activator protein), which binds to the lac promoter. CAP-promoter binding increases the affinity of RNA polymerase for the lac promoter, thus promoting transcription and activating lactose metabolism.

Gene Expression in Eukaryotes: How are Eukaryotic Genes Regulated?

GENOMIC CONTROL

Most Common Regulatory Changes:

DNA METHYLATION: *Example*: methylation of cytosines in certain 5'-CG-3' sequences turns off genes

HISTONE ACETYLATION: acetylation of lysine residues in H3 and H4 reduces their affinity for DNA; histone acetylation promotes genes expression

---

[4] Lactose is the disaccharide galactose β(1,4) glucose (but you knew that!)

[5] cAMP again?!! Food for thought: one way to tie a number of chapters together is to list out the regulatory effects of cAMP in various types of cells and pathways.

TRANSCRIPTIONAL CONTROL IS HEAVILY INFLUENCED BY:

CHROMATIN STRUCTURE: heterochromatin (too condensed to do transcription) vs. euchromatin (less condensed; varying levels of transcription activity)

GENE REGULATORY PROTEINS: bind to DNA to activate or repress genes

Mechanisms of Gene Regulatory Proteins:

competitive DNA binding of transcription factor proteins masking the activation surface

direct interaction with (binding to) transcription factors

GENE REARRANGEMENTS regulate certain genes, and may be involved in cell differentiation. *Example*: rearrangement of antibody genes in B lymphocytes.

GENE AMPLIFICATION by repeated rounds of replication within the amplified region (It's like setting your CD player to automatically repeat.) This occurs when the need for specific gene products is unusually high. For example, during the early developmental stages of fertilized eggs, the huge demand for protein synthesis requires amplification of rRNA genes.

ALTERNATIVE RNA PROCESSING TO CONTROL GENE EXPRESSION:

ALTERNATIVE SPLICING - joining of different combinations of exons ultimately results in different proteins. Example: tissue-specific forms of α–tropomyosin, a structural protein produced in various tissues.

ALTERNATIVE SITES TO ATTACH POLY A TAILS

LENGTH OF POLY A TAILS: Longer tails give more stability to mRNA, resulting in increased opportunity for translation.

RNA EDITING: certain bases are chemically modified, deleted, or added. *Example*: Deaminating cytosine to produce uracil changes a CAA codon for glutamine into a UAA codon, which is a stop signal, producing a shorter version of the protein.

mRNA TRANSPORT CONTROL THROUGH NUCLEAR PORE COMPLEXES

NUCLEAR EXPORT SIGNALS: capping, association with or presence of specific proteins, CBP (cap-binding protein)

TRANSLATIONAL CONTROL: ALTERING PROTEIN SYNTHESIS

…allows eukaryotic cells to respond to various stimuli (e.g., heat shock, viral infections, and cell cycle phase changes)

COVALENT MODIFICATION of translation factors (nonribosomal proteins that aid translation) alters translation rate or enhances translation of specific mRNAs.

SIGNAL TRANSDUCTION: RESPONDING TO ENVIRONMENTAL SIGNALS BY ALTERING GENE EXPRESSION; INVOLVES SIGNAL MOLECULES

Gene expression changes are initiated by ligand binding to a receptor (cell surface or intracellular).

Complicating features of intracellular signal transduction mechanisms include:

1. Each type of signal may activate one or more pathways.
2. Signal transduction pathways may converge or diverge.

Checkpoints in cell cycle phases prevent the cell from entering the next phase until conditions are optimal and specific signals are received. The mechanism of progression: alternating synthesis and degradation of cyclins, a group of regulatory proteins that bind to and activate cyclin-dependent protein kinases (Cdks). Cdks phosphorylate a variety of proteins that signal the cell past a checkpoint to the next phase of mitosis.

REGULATION OF CELL DIVISION:

Positive control: by binding growth factors to specialized cell receptors

Negative control: tumor suppressor genes (*Examples*: Rb gene and p53 gene); Apoptosis (programmed cell death) occurs if too much DNA damage has occurred and/or if DNA repair mechanisms are incomplete.

Binding of growth factors to cell surface receptors initiates a cascade of reactions that induces two classes of genes:

1. Early response genes are rapidly activated. *Example*: protooncogenes - normal genes that, if mutated, can promote carcinogenesis
2. Delayed response genes are induced by activities of transcription factors and other proteins that were produced or activated during the early response phase. Examples of delayed response gene products: Cdks, cyclins, and other components needed for cell division.

CHAPTER 18: ANSWERS TO EVEN-NUMBERED REVIEW QUESTIONS

2. Negative supercoiling facilitates the unzipping of DNA during the initiation phase of replication.

4. Briefly, prokaryotic DNA replication consists of DNA unwinding, RNA primer formation, DNA synthesis catalyzed by DNA polymerase and the joining of Okazaki fragments by DNA ligase. Prokaryotic DNA replication differs from the eukaryotic process in that prokaryotic replication is faster, and in prokaryotes the Okazaki fragments are longer.

6.

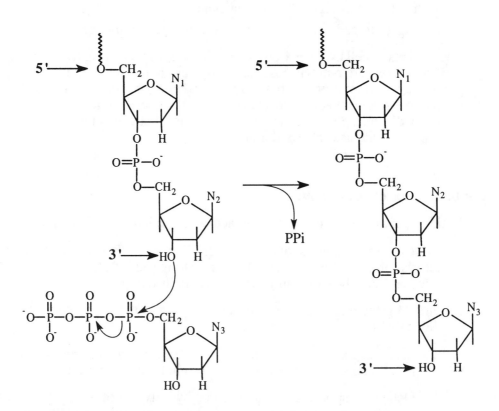

8. Viruses can cause mutations that affect the expression of protooncogenes by inserting their genomes into host cell regulatory sequences, thereby inactivating them.

10. Genetic recombination promotes species diversity. General recombination, a process in which segments of homologous DNA molecules are exchanged, is most commonly observed during meiosis. In site-specific recombination, protein-DNA interactions promote the recombination of nonhomologous DNA. Transposition is an example of site-specific recombination.

12. Intermicrobial DNA transfer mechanisms include transformation, transduction, and conjugation. In transformation, DNA fragments that enter the bacterial cell through cell wall openings and through a recombination event are inserted into the chromosome or a plasmid. In transduction, fragments of bacterial DNA are transferred to a recipient cell by a virus. Transduced DNA may insert into recipient

cell DNA through recombination. In conjugation, a donor cell produces a sex pilus that allows DNA to transfer to a recipient cell.

14. In DNA, if cytosine is converted to uracil, which forms a base pair with adenine, an AT base pair is substituted for a GC base pair. Such a change in RNA is not as important because RNA molecules are short lived and disposable. In contrast, because DNA molecules are the cell's permanent repository of genetic information, any change in base sequence may affect an organism's viability.

16. In both cases DNA copies are produced. However, in DNA replication usually only one copy of each DNA molecule is synthesized and several proofreading mechanisms ensure accurate copying. PCR technology is designed to produce multiple copies of a DNA molecule and proofreading is limited to the DNA polymerase that is employed.

18. a. DNA ligase – an enzyme that links DNA segments; responsible for linkage of Okazaki fragment during DNA synthesis
    b. DNA polymerase III – the major DNA synthesizing enzyme in prokaryotes
    c. SSB proteins – DNA binding proteins in prokaryotes that keep DNA strands separated during DNA synthesis
    d. primase – an RNA polymerase that synthesizes short RNA segments, called primers, that are required in DNA synthesis
    e. helicase – an ATP-requiring enzyme that catalyzes the unwinding of duplex DNA during DNA synthesis
    f. RPA – replication protein A; the primary homologue of SSB in eukaryotes
    g. Okazaki fragments – short DNA strands that are synthesized during discontinuous replication of the lagging DNA strand as the leading strand is continuously replicated

20. The processing steps that prepare a typical mRNA for its functional role include: capping (the linkage of a 7-methylguanosine to the 5'-end), cleavage of mRNA and addition of a poly A tail to the 3'-end, and splicing (the removal of introns).

22. The amplification of a single DNA molecule during 5 cycles yields 2^5 or 32 molecules.

CHAPTER 18: ANSWERS TO EVEN-NUMBERED THOUGHT QUESTIONS

2. DNA synthesis of the lagging strands occurs in the $5' \rightarrow 3'$ direction in a series of small pieces that are later joined together by DNA ligase.

4. Genetic recombination can allow cells to alter gene expression. The best known example is antibody production in lymphocytes. The rearrangement of several possible choices for each of a number of antibody gene segments via site specific recombination results in the generation of an extremely large number of different antibody molecules.

6. Any exposure to UV light causes genetic damage in skin cells. The production of melanin, the "tanning" substance that absorbs the energy of UV light, is a response to damage that has already occurred. This damage, which accumulates over years of exposure to UV light, accelerates the aging process, hence the wrinkled and

thickened skin. In some genetically predisposed individuals, the accumulating damage results in skin cancer.

8. Antibiotic resistance arises because the overuse of antibiotics acts as a selection pressure, i.e., they provide a growth advantage for disease-causing organisms that possess resistant genes. So-called superbugs are organisms that are resistant to several types of antibiotics because they possess plasmids containing several resistant genes. If the circumstances that cause antibiotic resistance continue, antibiotics may eventually become ineffective against most infectious diseases.

10. Errors that occur during DNA replication have the potential to become permanent if repair processes fail. Errors made during transcription affect only one or a few molecules and are temporary.

12. Refer to Biochemical Methods 18.1.

CHAPTER NINETEEN

Protein Synthesis

TRANSLATION of the nucleotide-base-sequence code results in a sequence of amino acids that form a polypeptide.

Protein synthesis also includes posttranslational modification and targeting.

THE GENETIC CODE: A CODING DICTIONARY THAT SPECIFIES A MEANING FOR EACH BASE SEQUENCE

CODON - a three-base-sequence (triplet) in mRNA that codes for each amino acid. For example, "CCC" codes for proline.

Of the 64 triplets that are possible, 61 code for amino acids and 3 are stop signals that terminate the growing polypeptide chain. (Stop codons: UAA, UGA, UAG)

PROPERTIES OF THE GENETIC CODE:

1. **DEGENERATE:** many codons have the same meaning; that is, most amino acids are specified by more than one codon. (Only Met and Trp have one codon.)

2. **SPECIFIC:** Each codon signals a specific amino acid. *Exception*: AUG codes for both Met and a start signal (initiating codon). Also, many of the codons that code for a specific amino acid have similar sequences. *Example*: UC? codes for Ser, whether the third base is U, C, A, or G.

3. **NONOVERLAPPING AND WITHOUT PUNCTUATION:** The code reads from the initiating codon (AUG) straight through to a stop codon, without repeating or skipping any bases.
   Reading Frame: set of side-by-side codons in mRNA

   Open reading frame: series of codons without a stop codon

   Why are reading frames important? Since the codons are sequential and nonoverlapping, the actual amino acid sequence depends on the starting point for the first triplet. Each different starting point defines a unique potential protein. Changing the starting point changes the reading frame, which changes the final product - the amino acid sequence. Check out the effects of changing the reading frame for the following sequence:

   5'- A G G C A G A A C U A A C C A G G U C U A - 3'

Frame 1:	AGG	CAG	AAC	UAA	CCA	GGU	CUA
	Arg	Gln	Asn	Stop			

Frame 2:	A GGC	AGA	ACU	AAC	CAG	GUC	UA
	Gly	Arg	Thr	Ile	Gln	Val	

Frame 3:	AG GCA	GAA	CUA	ACC	AGG	UCU	A
	Ala	Glu	Leu	Thr	Arg	Ser	

4. UNIVERSAL: Codons have the same meaning in different species.[1]

Codon-Anticodon Interactions

ANTICODON: tRNA's three-base sequence that base-pairs with its complementary codon on mRNA. tRNA has an anticodon on one side and a specific amino acid on the other.

Careful: Codon-anticodon pairings are antiparallel, but the sequences are always written in the 5'→ 3' direction. For example, the codon 5'-AGG-3' pairs with the anticodon 3'-UCC-5', but that anticodon would be written as CCU to be 5'→3'.

THE WOBBLE HYPOTHESIS: ALLOWS FOR MULTIPLE CODON-ANTICODON INTERACTIONS BY INDIVIDUAL TRNAS AND IS BASED PRINCIPALLY ON THESE OBSERVATIONS:

1. The first two base pairings in a codon-anticodon interaction confer most of the specificity required during translation.

2. The interactions between the third codon and anticodon nucleotides are less stringent, and non-traditional base pairs occur. That third anticodon nucleotide (in the 5'-position) corresponding to the third codon nucleotide in the mRNA sequence (in the 3'-position) is the "wobble" position.

The Aminoacyl-tRNA Synthetase Reaction: The Second Genetic Code

Aminoacyl-tRNA synthetases catalyze attachment of amino acids to tRNAs. This reaction is irreversible (due to PP_i hydrolysis). "The Second Genetic Code" reflects the importance of this reaction's accuracy.

amino acid + ATP + tRNA → aminoacyl-tRNA + AMP + PP_i .

There's at least one aminoacyl-tRNA synthetase for each amino acid. Many synthetases have a proofreading site to correct mistakes. For example, if isoleucyl-tRNA[Ile] synthetase makes Val–tRNA, its proofreading site will fit the Val end (but not Ile) and hydrolyze the incorrect bond.

LINKING AN AMINO ACID TO THE 3' TERMINUS OF THE CORRECT TRNA:

1. ACTIVATION OF THE AMINO ACID:

amino acid + ATP → aminoacyl-AMP + PP_i

---

[1] Thanks to MK for pointing out that in the strictest sense of the term, the genetic code isn't truly universal. For example, mitochondrial DNA has its own code that differs considerably from the general genetic code.

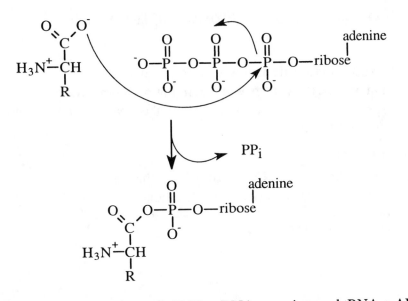

2. **tRNA LINKAGE:** aminoacyl-AMP + tRNA → amino acyl tRNA + AMP
   Linkage may be through the 2'-OH or the 3'-OH, but always happens at the 3'-end. The aminoacyl group can move between the 2' and 3' positions. Only the 3'-aminoacyl esters are used in protein synthesis.

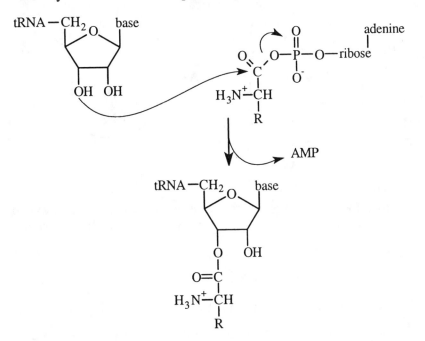

PROTEIN SYNTHESIS

Polysome: An mRNA with several ribosomes bound to it. One mRNA can be read at the same time by several ribosomes.

Two ribosomal sites for codon-anticodon interactions: P (peptidyl) and A (acyl)

Note that polypeptide synthesis proceeds from the N-terminal to the C-terminal because the message is read in the 5'→3' direction.

Requires GTP as an energy source, plus protein factors

221

PHASES OF TRANSLATION: INITIATION, ELONGATION, TERMINATION

INITIATION:

1. the small ribosomal subunit binds to an mRNA
2. initiator tRNA base pairs with the initiation codon AUG
3. large ribosomal subunit combines with the small subunit; the initiator tRNA is bound to the P site

ELONGATION CYCLE:

1. a second aminoacyl-tRNA base-pairs to the A site
2. peptidyl transferase catalyzes peptide bond formation (transpeptidation) - α-amino group of the A site amino acid attacks the C=O of the P site amino acid. Now both amino acids are on the A site tRNA.

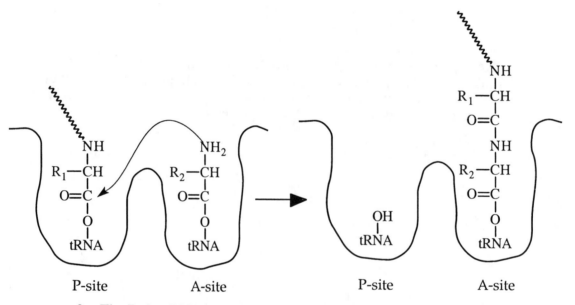

P-site   A-site   P-site   A-site

3. The P site tRNA leaves.
4. translocation - ribosome moves along the mRNA. The next codon enters the A site, and the tRNA with the growing peptide chain moves to the P site.

TERMINATION:

1. stop codon enters the A site.

2. protein releasing factor binds to the A site

3. peptidyl transferase hydrolyzes the polypeptide–tRNA ester bond, releasing the polypeptide chain

4. the ribosome releases mRNA and dissociates into its subunits

POSTTRANSLATIONAL MODIFICATIONS; PURPOSES:

1. to prepare a polypeptide for its specific function

2. targeting - to direct a polypeptide to a specific location

Prokaryotic Protein Synthesis

PROKARYOTIC RIBOSOMES are 70S in size and composed of a 50S (large) subunit (that contains catalytic site for peptide bond formation) and a 30S (small) subunit (that serves as a guide for regulatory translation factors).

IF = Initiation factor, EF = Elongation factor, RF = Release factor

The SHINE-DALGARNO SEQUENCE on the mRNA is located upstream from the initiation codon AUG. It binds to a complementary sequence on the small ribosomal subunit, and distinguishes AUG as a start codon from AUG as a methionine codon. Each gene has its own start codon and its own Shine-Dalgarno sequence.

The initiating tRNA is N-formylmethionine-tRNA or fmet-tRNA^{fmet}. The formyl group is added to This also helps to distinguish the start amino acid from an internal methionine.

N-formyl methionine is synthesized on the charged tRNA.

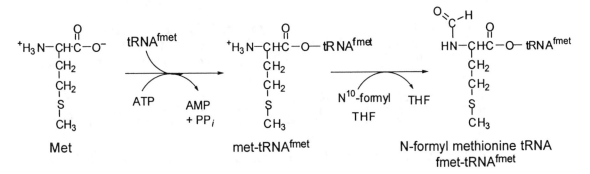

Met   met-tRNA^{fmet} N-formyl methionine tRNA fmet-tRNA^{fmet}

INITIATION

1. IF-1 and IF-3 bind to the 30S subunit: IF-1 binds to the A site and blocks it; IF-3 prevents the 30S subunit from binding to the large subunit too soon.
2. Shine-Dalgarno sequence base-pairs to the 30S subunit, guiding the mRNA which also binds to the 30S subunit
3. IF-2 (with a bound GTP) binds to the 30S subunit, promotes binding of the initiating tRNA, fmet-tRNA^{fmet}
4. fmet-tRNA^{fmet} binds to the initiating codon (AUG) on mRNA

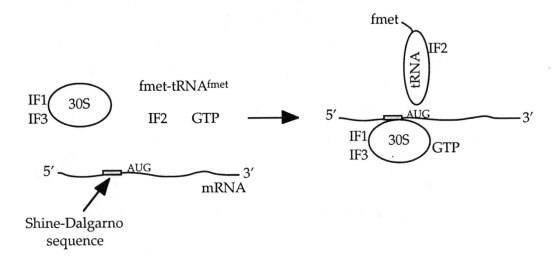

5. GTP hydrolyzes to GDP and PP$_i$, and the large and small subunits join.

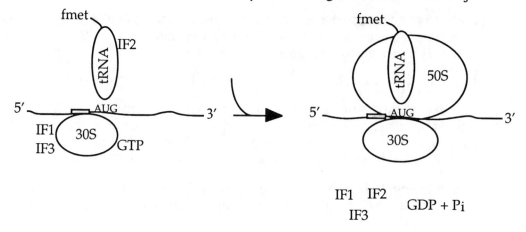

ELONGATION CYCLE:

1. The aminoacyl-tRNA needs to enter the empty A-site of the ribosome complex. An aminoacyl-tRNA binds *first* to EF-Tu-GTP, *then* to the A site; GTP hydrolysis releases EF-Tu-GDP (and a P$_i$) from the ribosome.

   EF-Ts regenerates EF-Tu-GTP:
   EF-Ts + EF-Tu-GDP → EF-Ts/EF-Tu-GDP → EF-Ts/EF-Tu + GDP

   EF-Ts/EF-Tu + GTP → EF-Ts/EF-Tu-GTP → EF-Ts + EF-Tu-GTP

2. Peptidyl transferase catalyzes peptide bond formation. Now both amino acids are on the A site tRNA, and the tRNA on the P site leaves.
3. Translocation: The polypeptide-tRNA moves from the A-site to the P-site, and the ribosome moves so that the next codon on the mRNA is in the A-site. This stage requires EF-G and GTP hydrolysis.

The elongation process repeats until the ribosome reaches a stop codon.

TERMINATION REQUIRES RELEASING FACTORS RF-1, RF-2, AND RF-3

There are no tRNAs for stop codons (UAA, UGA, UAG). Instead, releasing factors (RFs) are used. RF-1 recognizes UAA and UAG, and RF-2 recognizes UAA and UGA. (RF-3 may promote RF-1 and RF-2 binding.) A GTP is hydrolyzed, and peptidyl transferase hydrolyzes the polypeptide-tRNA bond, releasing the polypeptide. The mRNA and tRNA dissociates from the ribosome, which separates into its subunits.

4. POSTTRANSLATIONAL MODIFICATIONS

1. Proteolytic processing: cleavage reactions, including the removal of formylmethionine and signal peptide sequences. Signal peptides (leader peptides) are short peptides, typically near the N-terminus, that determine a polypeptide's destination.
2. Conjugation: typically to form lipoproteins, some instances of glycosylation (covalently linking carbohydrates to proteins) are also known

3. Methylation: Protein methyltransferases use SAM to add methyl groups (reversibly) to amino acid residues of proteins; plays a role in signal transduction, for example.

4. Phosphorylation/dephosphorylation: catalyzed by protein kinases and phosphatases; plays a role in chemotaxis and nitrogen metabolism regulation

TRANSLATIONAL CONTROL MECHANISMS

Protein synthesis in prokaryotes is controlled predominantly by the rate of transcription, although several types of translational regulation have been detected: 1) differences in Shine-Dalgarno sequences, 2) negative translational regulation: *Example*: Some ribosomal proteins inhibit the translation of their own or related operons.

Eukaryotic Protein Synthesis

EUKARYOTIC RIBOSOMES are 80S in size, with a 40S small subunit and a 60S large subunit. The number and size of rRNAs and ribosomal proteins also differ.

RATE OF SYNTHESIS: Prokaryotic translation proceeds at a rate of 1,200 amino acids per minute, whereas eukaryotic protein synthesis is much slower at about fifty amino acids per minute.

eIF, eEf, eRf: eukaryotic initiation, elongation, and releasing factors

Features of Eukaryotic Translation:

INITIATION:

On the road to the final 80S initiation complex [40S-60S-mRNA-met-tRNA$_i$], an ATP and a GTP are hydrolyzed, a 40S preinitiation complex and a 40S initiation complex are formed, and at least nine eIFs are used.

Eukaryotic initiation is much more complex than prokaryotic initiation because:

1. **mRNA SECONDARY STRUCTURE:** Eukaryotic mRNA has a cap and a tail, and has had its introns removed. Also, mRNA, made inside the nucleus, might interact with cellular proteins before it gets to the ribosome that's outside of the nucleus. (ribonucleoprotein particle = mRNA + protein)

2. **mRNA SCANNING:** Ribosomes bind to the capped 5' end and move towards the 3' end, searching for a translation start site. This is needed because eukaryotic mRNA lack Shine-Dalgarno sequences.

ELONGATION: Eukaryotic elongation is very similar to prokaryotic elongation.

eEF-1α is analogous to EF-Tu. eEF-1β and eEF-1γ work together to regenerate eEF-1α-GTP from eEF-1α-GDP. (This is similar to the action of EF-Ts). In translocation, eEF-2 is analogous to EF-G.

Kinetic proofreading: if correct pairing doesn't occur between the A site and the eEF-1α-GTP-aminoacyl-tRNA, the complex leaves the A site.

TERMINATION: Eukaryotic termination is also very similar to prokaryotic termination.

GTP binds to eRF-3, which then forms a complex with eRF-1. This complex binds in the A site when a stop codon (UAG, UGA, or UAA) enters.

GTP hydrolysis promotes dissociation of the eRFs from the ribosome; and peptidyl transferase hydrolyzes the polypeptide-tRNA bond, releasing the polypeptide. The mRNA and tRNA dissociates from the ribosome, which separates into its subunits.

EUKARYOTIC POSTTRANSLATIONAL MODIFICATIONS INCREASE THE STRUCTURAL AND FUNCTIONAL DIVERSITY OF PROTEINS.

1. Proteolytic Cleavage: Proteases hydrolyze specific peptide bonds.

   Purpose: a common regulatory mechanism

   Examples: removing N-terminal Met and signal peptides; conversion of inactive proproteins to their active forms (proinsulin → insulin; zymogens or proenzymes → enzymes) (preproprotein = proprotein + signal peptide)

2. Glycosylation: adding sugar groups (secreted proteins have complex oligosaccharides, ER membrane proteins have high mannose species)

3. Hydroxylation of the amino acids proline and lysine by prolyl-4-hydroxylase, prolyl-3- hydroxylase, and lysyl hydroxylase in the RER; substrate requirements are highly specific. *Examples*: Hydroxylated proline and lysine are required for the structural integrity of collagen and elastin. 4-hydroxyproline is found in acetylcholinesterase and complement

4. Phosphorylation: metabolic control and signal transduction

5. Lipophilic modifications: Covalent attachment of lipid moieties to proteins enhances membrane binding capacity and/or certain protein-protein interactions. Most common: acylation (attaches a fatty acid) and prenylation

6. Methylation by methyltransferases may alter the cellular roles of certain proteins. Methylation of altered Asp residues promotes either the repair or the degradation of damaged proteins. Other examples of amino acid methylation: Lys, His, Arg

7. Disulfide bond (S–S) formation: in secreted proteins and certain membrane proteins, because reducing agents (like GSH) in the cytoplasm will reduce –S–S– bonds to –SH + HS–.

   Disulfide exchange - disulfide bonds rapidly migrate from one position to another until the most stable structure is achieved. (S–S bonds are not necessarily formed sequentially.)

8. Protein splicing: An intein (internal section of a protein) is cut out of a protein, and the two flanking sections - the exteins - are spliced together. Protein splicing is self-catalyzed, i.e., it requires no other enzymes, cofactors, or energy sources.

TARGETING DIRECTS THE PROTEIN TO ITS PROPER DESTINATION

Targeting mechanisms:

TRANSCRIPT LOCALIZATION: A specific mRNA binds to receptors in certain cytoplasmic locations. Translation of this localized mRNA results in a cytoplasmic protein gradient, that is, an asymmetrical distribution of this protein in the cytoplasm.

SIGNAL PEPTIDES are sorting signals that target polypeptides to their proper location (for secretion, or for use in the plasma membrane or any of the membranous organelles). Signal peptides help to insert the polypeptide that contains it into an appropriate membrane.

THE SIGNAL HYPOTHESIS

The signal hypothesis was proposed to explain how polypeptides translocate across RER membrane. Its significance is that it helps explain the ability of proteins to be specifically targeted to their proper location. The fate of a targeted polypeptide depends on the location of the signal peptide and other signal sequences.

SIGNAL RECOGNITION PARTICLE (SRP) binds to a ribosome and interrupts translation. The SRP mediates binding of the ribosome to RER via a docking protein (SRP receptor protein). Translation restarts, and the growing polypeptide inserts into the membrane.

COTRANSLATIONAL TRANSFER - simultaneous translocation of a polypeptide during ongoing protein synthesis

TRANSLATIONAL TRANSLOCATION - integral membrane protein complex that mediates polypeptide translocation

POSTTRANSLATIONAL TRANSLOCATION - previously-synthesized polypeptides are pulled across the RER membrane by an ATP-binding peripheral translocation-associated protein (hsp70).

Typically, after a protein is in the RER, it undergoes initial posttranslational modifications, is transferred to the Golgi complex via transport vesicles that bud off from the ER and fuse with the *cis* face of the Golgi membrane. Further protein modifications are made inside the Golgi complex. Transport vesicles exit from the *trans* face of the Golgi and move to target locations.

TRANSLATION CONTROL MECHANISMS

Features of eukaryotic translational control:

1. **mRNA EXPORT:** The spatial separation of transcription and translation that is afforded by the nuclear membrane appears to provide eukaryotes with the opportunity to control translation. Export through the nuclear pore complex is known to be a carefully controlled, energy-driven process whose minimum requirements include a 5'-cap and a 3' poly A tail.

b. **MRNA** STABILITY = HOW WELL **MRNA** CAN AVOID DEGRADATION BY NUCLEASES
In general, the translation rate of any mRNA species is related to its abundance, which in turn is dependent on both its rates of synthesis and degradation. The length of the poly A tail is significant and affects its stability.

c. **NEGATIVE TRANSLATIONAL CONTROL:** The translation of some mRNAs is known to be controlled by the binding of repressor proteins to the 5' ends of the mRNA. This effectively blocks ribosome binding and scanning.

d. **INITIATION FACTOR PHOSPHORYLATION:** The phosphorylation of eIF-2 in response to certain stimuli (e.g., heat shock, viral infections, and growth factor deprivation) has been observed to decrease protein synthesis. However, the translation of certain mRNA increases (e.g., hsp synthesis in response to heat shock).

e. **TRANSLATIONAL FRAMESHIFTING:** This process, often observed in retroviruses-infected cells, allows the synthesis of more than one polypeptide from a single mRNA.

The Folding Problem

THE TRADITIONAL FOLDING MODEL: Interactions between amino acids side chains alone force the molecule to fold into its final shape.

LIMITATIONS OF THE TRADITIONAL FOLDING MODEL

1. Time constraints: Folding happens on the order of seconds (or a few minutes), not in years, as calculated by the traditional folding model.

2. Complexity: Think about the number of bonds that can rotate, not only in the backbone, but in the side groups. Phew!

MOLECULAR CHAPERONES AID PROTEIN FOLDING AND TARGETING IN LIVING CELLS

These chaperone proteins bind to denatured or unfolded proteins and assist in adoption of the correct three-dimensional structure. Most appear to be hsps (heat shock proteins). *Examples:*

1. Hsp70s - bind to short hydrophobic segments in unfolded polypeptides to prevent their aggregation. ATP hydrolysis releases the polypeptide, which is then passed on to an hsp60.

2. Hsp60s (chaparonins or Cpn 60s) mediate protein folding.

CHAPTER 19: ANSWERS TO EVEN-NUMBERED REVIEW QUESTIONS

2. The observations upon which the wobble hypothesis is based are: (1) the first two base pairings in a codon-anticodon interaction confer most of the specificity required during translation, and (2) the interactions between the third codon and anticodon nucleotides are less stringent. Because of the "wobble rules," only a minimum of 31 tRNAs are required for the translation of all 61 codons.

4. The major differences between prokaryotic and eukaryotic translation are speed (the prokaryotic process is significantly faster), location (the eukaryotic process is not directly coupled to transcription as prokaryotic translation is), complexity (because of their complex life styles, eukaryotes possess complex mechanisms for regulatory protein synthesis, e.g., eukaryotic translation involves a significantly larger number of protein factors than prokaryotic translation), and posttranslational modifications (eukaryotic reactions appear to be considerably more complex and varied than those observed in prokaryotes).

6. During the elongation phase of protein synthesis, the second aminoacyl-RNA becomes bound to the ribosome in the A site. Peptide bond formation is then catalyzed by peptidyl transferase. Subsequently, the ribosome is moved along the mRNA by a mechanism referred to as translocation.

8. a. targeting – a series of mechanisms that directs newly synthesized polypeptides to their correct cellular locations.

   b. scanning – a mechanism that eukaryotic ribosomes use to locate a translation start site on an mRNA.

   c. codon – an mRNA triplet base sequence that specifies the incorporation of a specific amino acid into a growing polypeptide chain during translation or acts as a start or stop signal.

   d. reading frame – a set of contiguous triplet codons.

   e. molecular chaperones – molecules that assist in the folding and targeting of proteins.

   f. disulfide exchange – a mechanism that facilitates the formation of disulfide bridges in newly synthesized proteins.

   g. proofreading site – a second active site of certain aminoacyl-tRNA synthetases which binds a specific tRNA if it is covalently bound to the wrong amino acid and hydrolyzes the tRNA-amino acid bond

   h. signal peptides – short peptide sequences that determine a polypeptide's destination, e.g., by directing its insertion into a membrane.

   i. glycosylation – a posttranslational mechanism whereby carbohydrate groups are covalently attached to polypeptides.

   j. negative translational regulation – the blockage of the translation of a specific mRNA when a specific protein binds to a sequence near its 5'-end

10. Cotranslational transfer is a process in which nascent polypeptides are inserted through an intracellular membrane during ongoing protein synthesis. An integral membrane protein complex referred to as the translocon, mediates the transfer of polypeptides (each of which contain some hydrophilic residues) across the hydrophobic core of the membrane.

12. Synthesis of a secretory glycoprotein begins on a ribosome. An appropriate signal peptide mediates the translocation of the polypeptide into the ER lumen. The core N-linked oligosaccharides are then covalently linked to appropriate asparagine residues in the polypeptide in a reaction catalyzed by glucosyl transferase. Subsequently, the molecule is transferred in transport vesicles to the Golgi complex where additional glycosylation reactions occur. Eventually, the glycoprotein is incorporated into secretory vesicles which migrate to the plasma membrane. Secretion of the glycoprotein then occurs via exocytosis.

14. The hsp70s are a family of ATP-binding molecular chaperones that bind to and stabilize protein during the early stages of protein folding. The hsp60s form a large structure composed of two stacked rings in which proteins fold in an ATP-dependent process.

16. In protein splicing an intervening peptide sequence is excised from a nascent polypeptide. A peptide bond is formed between the flanking amino terminal and carboxy terminal amino acid.

18. GTP hydrolysis provides the energy that drives the movement of the peptidyl-tRNA from the A site to the P site in the ribosome.

CHAPTER 19: ANSWERS TO EVEN-NUMBERED THOUGHT QUESTIONS

2. Because the accuracy of protein synthesis depends directly on codon-anticodon interactions, the specificity with which t-RNAs are linked to amino acids is critically important. The process in which the amino acid-tRNA synthetases catalyze the covalent binding of each of the t-RNAs with its correct amino acid has, therefore, been referred to as the second genetic code.

4. The three phases of protein synthesis are: (1) initiation (the small ribosomal subunit binds to an mRNA, an initiation tRNA and the large subunit), (2) elongation (polypeptide synthesis occurs as amino acids attached to tRNAs aligned in the P and A sites undergo transpeptidation), and (3) termination (polypeptide synthesis ends when a stop codon on the mRNA enters the A site). Translation factors perform a variety of roles. Some have catalytic functions (e.g., EF-TU and eEF-2 are GTP binding proteins which catalyze GTP hydrolysis), while others (e.g., IF-3 and eIF-4F) stabilize translation structures.

6. The phases of protein synthesis during which each process occurs are as follows: a. Initiation, b. Elongation, c. Elongation, and d. Termination.

8. The nucleotide GTP is the source of the energy required to drive various steps in the translation mechanism.

10. Each Shine-Dalgarno sequence in a prokaryotic mRNA occurs near a start codon (AUG). The Shine-Dalgarno sequence provides a mechanism for promoting the correct alignment of the start codon on the ribosome (as opposed to a methionine codon) because it binds to a nearby complementary sequence in the 16S rRNA component of the 30S ribosome. Eukaryotic ribosomes identify the initiating AUG codon by binding to the capped 5' end of the mRNA and scanning the molecules for a translation start site.

12. Sets of amino acids which may require proofreading include phenylalanine/tyrosine, serine/threonine, aspartate/glutamate, asparagine/glutamine, isoleucine/leucine, and glycine/alanine.

14. The principal factors that ensure accuracy in protein synthesis are codon-anticodon base pairing and the mechanism by which amino acids are linked to their cognate tRNAs. The level of accuracy of protein synthesis, while quite high, is still less than that achieved during replication or transcription.

16. In sickle cell anemia the β-chain has a valine instead of glutamate at position 6. Glutamate is coded for by GAA or GAG, and valine is coded for by GUU, GUC, GUA, or GUG. The mutation is caused by the replacement of uracil for adenine in the β–chain mRNA. This change in turn is created by a substitution of thymine for adenine in the original DNA.